Electrical Circuits and Systems

Noel M. Morris

Principal Lecturer,
North Staffordshire Polytechnic

Second Edition

First edition 1975
Second edition 1980
Reprinted 1981

Published by
THE MACMILLAN PRESS LTD
London and Basingstoke
Companies and representatives throughout the world

Typeset in 10/12 Times by
REPRODUCTION DRAWINGS LTD, Sutton, Surrey

Printed in Hong Kong

British Library Cataloguing in Publication Data

Morris, Noel Malcolm
    Electrical circuits and systems. - 2nd ed. -
    (Macmillan basis books in electronics).
    1. Electric circuits
    I. Title
    621.319'2   TK454

    ISBN 0-333-29115-8

Contents

Contents

Foreword

Technological progress has nowhere been more rapid than in the fields of electronics, electrical, and control engineering. The *Macmillan Basis Books in Electronics Series* of books have been written by authors who are specialists in these fields, and whose work enables them to bring technological developments sharply into focus.

Each book in the series deals with a single subject so that undergraduates, technicians, and mechanics alike will find information within the scope of their courses. The books have been carefully written and edited to allow each to be used for self-study; this feature makes them particularly attractive not only to readers approaching the subject for the first time, but also to mature readers wishing to update and revise their knowledge.

Noel M. Morris

Preface

This book deals with the principles of electrical circuits and systems, and to enable the reader to improve his understanding of the subject, a large number of worked examples are included in the text. Consequently, those attending a wide range of courses at universities, polytechnics, and colleges of further education will benefit from the book. It will also be of service to mature engineers wishing to refresh their knowledge. The book also provides other readers who are not directly concerned with electrical engineering with the principles of the subject.

The book begins with a chapter on the basis of electrical circuits, dealing with the fundamental concepts and units involved. There follow three chapters on important aspects of circuits, namely circuit theorems, electromagnetism, and electrostatics. After this, attention is directed to the solution of alternating current circuits — in four chapters the topics of alternating voltage and current, single-phase circuits, complex notation, and polyphase circuits are discussed. The transformer is the subject of a chapter in which it is considered variously as a power-transforming device, as an impedance level converting device, and as an element in a coupled circuit. The book concludes with a chapter on transient effects in electrical circuits.

I would like to acknowledge the encouragement I have received from my wife during the preparation of this book, and the forbearance of my family.

In this second edition the opportunity has been taken to improve the book by making a number of modifications to the text. Problems (with answers) have also been included at the end of each chapter; this will make the book even more useful to students of electronic and electrical engineering and other readers of the book. Such errors as have been brought to the author's attention have also been rectified.

Noel M. Morris

1 The Electrical Circuit

1.1 Conductors, Semiconductors, and Insulators

Materials used in electrical and electronic engineering can be classified in many ways, one of which is their ability to conduct electricity. Broadly speaking, most materials can be described as being conductors, or semiconductors, or insulators. *Conductors* are materials which readily allow current flow to occur when an e.m.f. is applied, whereas the current flow in *insulators* is very small (ideally, it is zero). *Semiconductors* are materials whose resistance to flow of current lies between that of conductors and that of insulators. A constant or *parameter* often used to describe the ability of a material to resist the flow of current is its *resistivity* (see section 1.10 for a full discussion). Using resistivity as a means of defining the above categories of material, conductors have a resistivity in the range from zero to 10^{-4} Ωm, semiconductors have a resistivity in the range 10^{-4} to 10^3 Ωm, while the resistivity of insulators is above 10^3 Ωm.

The ability of materials to conduct current depends on their chemical structure. In the following section we shall discuss the relevant features of the atomic structure of materials used in electrical and electronic engineering.

1.2 Structure of the Atom

Although it is physically impossible to 'see' an atom, we can use experimental techniques to measure the size, mass, and charge of the invisible particle. Our concept of the atom is continually evolving, but we know that its structure contains a *nucleus*, consisting of *protons* and *neutrons*, which is surrounded by orbiting *electrons*. The orbits are known as *shells* or *energy levels*. The mass and charge associated with electrons, protons, and neutrons are given in table 1.1. From the table we see that electrons and protons have equal but opposite charges, and that neutrons have no charge and are electrically neutral. On the other hand, the mass of the electron is much less than that of the proton, while the masses of protons and neutrons are equal. In addition to the basic atomic particles, atoms also contain a number of particles which are unstable in nature and quickly disappear. These unstable particles have no significance in electrical engineering.

T able 1.1 Basic atomic particles

Particle	Mass (kg)	Charge (C)
Electron	9.11×10^{-31}	-0.16×10^{-18}
Proton	1.67×10^{-27}	$+0.16 \times 10^{-18}$
Neutron	1.67×10^{-27}	nil

The nucleus

The nucleus contains protons and neutrons, and has a net positive charge equal to that of the protons. There are *no* electrons in the nucleus. The *atomic number* of that atom is equal to the number of protons in the nucleus.

Electrons

The net electrical charge of an isolated atom is zero, and it follows that there are as many electrons in orbit around the nucleus as there are protons in the nucleus. Thus the copper atom, which has 29 protons in its nucleus has 29 electrons in orbit. Also, the electrons orbit in particular layers or shells around the nucleus and the higher the electronic energy level of the electron, the greater its radius of orbit. The shells have been designated letters by scientists, commencing with K for the innermost shell. Successive shells are called the L, M, N, O, P, and Q shells. It has also been found that all shells except the K shell comprise a number of subshells, which allow the electrons to assume one of several types of orbit. For example, electrons in one subshell may follow a circular orbit, and those in another subshell may follow an elliptical orbit. These variations allow the magnetic properties of some substances to be explained.

The valence shell

In an isolated atom, the valence energy shell is the highest energy level at which electrons are found at absolute zero temperature. As the temperature of the atom rises, the *valence electrons* (the electrons in the valence shell) become more 'energetic' and the additional energy causes some of them to transfer to a higher energy level. To do so, the electrons must 'leap' across the energy gap between the valence shell and the next higher available shell, which is known as the *conduction shell.* Electrons which manage to transfer across the gap, known as the *forbidden energy gap,* can then take part in electrical conduction.

   A simple analogy of energy levels in an atom is presented in the form of a pan of water that is being heated. The drops of water on the upper surface level correspond to the electrons in the valence shell. As the pan is heated some of the drops of water on the upper surface acquire sufficient energy to escape to the atmosphere, which is equivalent in the atom to electrons transferring to the conduction energy band.

As isolated atoms are brought closer together to form a solid, the particles condense from the gaseous state to the solid state, when the energy shells spread to form *energy bands*, the bands being separated once more by forbidden energy gaps. Thus, in a solid the valence electrons reside in a *valence energy band*, the next available band above it being the *conduction band*. Bringing the atoms even closer together causes the valence and conduction bands to touch or even overlap.

1.3 Classification of Solids

Materials used in electrical and electronic engineering can be classified according to the *energy-band theory* outlined in section 1.2. That is, the materials have either

(1) a full valence band which is separated from the conduction band by an energy gap, or
(2) a full or partially full valence band which overlaps the conduction band.

Insulators and semiconductors are contained in the first group, while the latter group contains electrical conductors. The feature that distinguishes insulators from semiconductors is the size of the energy gap between the valence and conduction bands. In insulators the gap is large so that very few electrons manage to traverse it at room temperature, whereas in semiconductors the gap is much smaller.

1.4 Thermal Effects on the Electrical Resistance of Insulators and Semiconductors

The electrical resistance to current flow in solids varies with temperature, and the reason for the way in which the resistance changes is suggested by the energy-band theory of solids.

As we have seen, when the temperature of materials with an energy gap between the valence and conduction bands is increased, more electrons can transfer into the conduction band. This causes the electrical resistance of these materials to decrease with increase in temperature.

1.5 Basic Electrical Quantities

In this chapter we are concerned with the flow of *direct current* (d.c.) in electrical circuits. The following are terms and units used in association with these circuits.

Electrical current

Symbol I. The **ampere** (A) is that current which, when flowing in each of two infinitely long parallel conductors of negligible cross-section, and placed 1 m apart in a vacuum, produces between the conductors a force of 2×10^{-7} N per metre length.

Electrical quantity

Symbol Q. The **coulomb** (C) is the quantity of electricity passing a point in a circuit when a current of 1 ampere flows for 1 second, and

$$Q = It \quad \text{coulombs}$$

Electrical potential

Symbol E. The unit of electric potential is the **volt** (V), and is the potential difference that exists between two points on an electrical conductor which carries a current of 1 ampere, the electrical resistance between the two points being 1 ohm.

$$E = IR \quad \text{volts} \qquad\qquad (1.1)$$

The relationship given in equation 1.1 is known as **Ohm's law**.

Electrical resistance

Symbol R. The unit of electrical resistance is the **ohm** (Ω) and, when a current of 1 A flows through a conductor of resistance 1 Ω, it causes the p.d. between the ends of the conductor to be 1 V.

Electrical energy

Symbol W. The **joule** (J) is the energy dissipated in a circuit when a p.d. of 1 V causes a current of 1 A to flow for 1 second.

$$W = EIt \quad \text{joules or watt seconds}$$

Since a 100 W lamp consumes in one hour a total of $100 \times 60 \times 60 = 360\,000$ J, we see that the joule is not a practical unit of energy. The commercial unit of electrical energy is the kilowatt hour (1000 watt hour), which is

$$1 \text{ kWh} = 1000 \times 60 \times 60 \text{ J} = 3\,600\,000 \text{ J or } 3.6 \text{ MJ}$$

The kWh is often called a *unit* of electrical energy, and

$$\text{kWh} = \frac{\text{joules}}{3.6 \times 10^6}$$

Electrical power

Symbol P. Power is the rate of doing work, and the unit is the watt (**W**) or joule/second.

$$P = \frac{EIt \text{ joules}}{t \text{ seconds}} = EI \quad \text{watts}$$

1.6 Relationships in an Electrical Circuit

The following relationships were stated in section 1.5.

$$E = IR \quad \text{V} \quad \textbf{(Ohm's law)} \qquad (1.2)$$

$$Q = It \quad \text{C} \qquad (1.3)$$

$$P = EI \quad \text{W} \quad {}^{= I^2 R.} \qquad (1.4)$$

$$W = EIt \quad \text{J} \quad {}^{Et}\!/\!_R \qquad (1.5)$$

Using the above, several other important relationships can be deduced. For example, if we substitute equation 1.2 into equation 1.4 we have

$$P = EI = (IR)I = I^2R \quad \text{W} \qquad (1.6)$$

Also, substituting $I = E/R$ from equation 1.2 into equation 1.4 gives

$$P = EI = E \times \frac{E}{R} = \frac{E^2}{R} \quad \text{W} \qquad (1.7)$$

Substituting equations 1.6 and 1.7 into equation 1.5 gives

$$W = I^2Rt \quad \text{J} \qquad (1.8)$$

$$= \frac{E^2t}{R} \quad \text{J} \qquad (1.9)$$

Example 1.1

An electrical circuit has a resistance of 200 Ω and is energised by a 250 V supply. Calculate the current drawn from the supply.

Solution

$$I = \frac{E}{R} = \frac{250}{200} = 1.25 \text{ A}$$

Example 1.2

Calculate the power dissipated by the circuit in example 1.1.

Solution

$$P = EI = 250 \times 1.25 = 312.5 \text{ W}$$

Note This can also be calculated from either of equations 1.6 and 1.7.

Example 1.3

Determine the electrical energy consumed in 1 hour by the circuit in example 1.1.

Solution

$$W = EIt = 250 \times 1.25 \times 60 \times 60 = 1\ 125\ 000 \quad \text{J}$$

$$= \frac{1\ 125\ 000}{3.6 \times 10^6}\ \text{kWh} = 0.3125\ \text{kWh}$$

Note This can also be calculated from either of equations 1.8 and 1.9.

Example 1.4

Calculate the quantity of electricity consumed by the circuit in example 1.1 in a period of 60 seconds.

Solution

$$Q = It = 1.25 \times 60 = 75\ \text{C}$$

1.7 Multiples and Sub-multiples of Electrical Quantities

Many of the basic units used in electrical circuits are, in some applications, either too large or too small for use with practical circuits. For example, the basic unit of power (the watt) is too small to describe the power output from an electrical generator; we rate these devices either in terms of the number of kilowatts (1000s of watts) or of the number of megawatts (1 000 000s of watts) generated. The power dissipated by transistors is very small and, in this case, the watt is an inconveniently large unit. Here we use the milliwatt (10^{-3} W) as the basic unit of power. The multiples and sub-multiples used in connection with electrical circuits are listed in table 1.2.

Example 1.5

The insulation resistance of a cable is 10 MΩ. Calculate the leakage current which flows through the cable insulation when the supply voltage is 200 V.

Solution

$$I = \frac{E}{R} = \frac{200}{10 \times 10^6} = 20 \times 10^{-6}\ \text{A} = 20\ \mu\text{A}$$

Table 1.2

Symbol	Prefix	Multiple
T	tera	10^{12}
G	giga	10^{9}
M	mega	10^{6}
k	kilo	10^{3}
c	centi	10^{-2}
m	milli	10^{-3}
μ	micro	10^{-6}
n	nano	10^{-9}
p	pico	10^{-12}
f	femto	10^{-15}
a	atto	10^{-18}

Example 1.6

Calculate the power dissipated in the insulation in example 1.5.

Solution

$$P = EI = 200 \times 20 \times 10^{-6} = 4000 \times 10^{-6} \text{ W} = 4000 \ \mu\text{W}$$
$$= 4 \text{ mW}$$

Example 1.7

Calculate the energy consumed in the insulation in example 1.6 if the supply is maintained for 60 seconds.

Solution

$$W = EIt = 200 \times 20 \times 10^{-6} \times 60$$
$$= 0.24 \text{ J} = 240 \text{ mJ}$$

1.8 Resistance Colour-coding and Preferred Values

The values of resistors used in many electronic circuits are colour-coded according to an international notation which is given in table 1.3; the coding method for resistors with axial leads is shown in figure 1.1.

Table 1.3

Colour	Significant figure	Decimal multiplier	Tolerance (per cent)
no band			20
silver		0.01	10
gold		0.1	5
black	0	1	
brown	1	10	
red	2	10^2	
orange	3	10^3	
yellow	4	10^4	
green	5	10^5	
blue	6	10^6	
violet	7	10^7	
grey	8	10^8	
white	9	10^9	

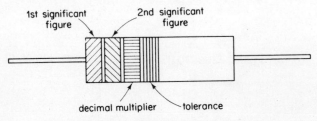

Figure 1.1   Resistor colour code

Thus a resistor that is colour-coded yellow, violet, orange, silver is a 47 kΩ resistor that has a tolerance of 10 per cent. That is, its value lies in the range 47 kΩ ± 10 per cent, or 42.3 kΩ to 51.7 kΩ. The nominal values of resistance that are commercially available are selected so that the value of a resistance at the upper limit of its tolerance band is approximately equal to the value of the next higher resistance at the lower limit of its tolerance. A list of these nominal values or **preferred values**, together with their tolerances are given in table 1.4. The values in use are decimal multiples and sub-multiples of those listed.

A convenient method of remembering the above colour code is given in the following mnemonic (the first letter of successive words in the mnemonic corresponds to the first letter in successive colours in the colour code):

Bye Bye Rosie Off You Go, Bristol Via Great Western

Table 1.4

Percentage tolerance		
20 per cent	10 per cent	5 per cent
10	10	10
		11
	12	12
		13
15	15	15
		16
	18	18
		20
22	22	22
		24
	27	27
		30
33	33	33
		36
	39	39
		43
47	47	47
		51
	56	56
		62
68	68	68
		75
	82	82
		91

1.9 Conductance

Conductance, symbol G, is the reciprocal of resistance, that is, $G = 1/R$, and has the unit of the **siemens** (S). Thus, a resistance of 10 Ω has a conductance of 0.1 S. Substituting the above relationship into Ohm's law we get

$$E = IR = \frac{I}{G} \quad \text{V}$$

or

$$I = EG \quad \text{A} \tag{1.10}$$

1.10 Resistivity and Conductivity

The resistance of a conductor depends not only on the physical dimensions of the conductor, but also on a parameter known as its resistivity. Experiments on a uniform sample of conducting material show that the resistance is directly proportional to its length, *l*, in the direction of flow of current, and inversely proportional to the area, *a*, through which the current flows. The **resistivity**, symbol ρ, relates the resistance of the sample to *l* and *a*, as shown below.

$$R = \rho \frac{l}{a} \tag{1.11}$$

In the SI system of units, *l* has the dimensions of metres, *a* has the dimensions of $(\text{metres})^2$, *R* is in ohms, and ρ has the dimensions of **ohm metres** (Ω m).

Typical values of resistivity are, for copper and aluminium at $20°C$, 1.73×10^{-8} Ω m (that is, 1.73×10^{-2} $\mu\Omega$ m or 17.3 $\mu\Omega$ mm) and 2.83×10^{-8} Ω m, respectively.

Conductivity

Symbol σ, is the reciprocal of resistivity, as follows.

$$\sigma = \frac{1}{\rho} \quad (\Omega m)^{-1} \tag{1.12}$$

hence

$$R = \frac{l}{\sigma a} \quad \Omega \tag{1.13}$$

or

$$G = \sigma \frac{a}{l} \quad S \tag{1.14}$$

Example 1.8

Determine the resistance of 100 m of copper wire of area 0.05 cm^2, the resistivity of the copper being 1.73×10^{-8} Ω m.

Solution

$$\text{Area} = 0.05 \times (10^{-2})^2 = 0.05 \times 10^{-4} \text{ m}^2$$

$$R = \rho \frac{l}{a} = 1.73 \times 10^{-8} \times \frac{100}{0.05 \times 10^{-4}}$$

$$= 0.346 \ \Omega$$

1.11 Temperature Coefficient of Resistance

The resistance of the majority of materials used in electrical engineering alters with change in temperature. In conductors this change is an increase in resistance for an increase in temperature, and in insulators and semiconductors it is a decrease in resistance for an increase in temperature (see section 1.4). Certain alloys such as constantan, eureka, and manganin show very little change in resistance over their working range. The latter group of materials is used in the construction of precision resistors and instrument shunts.

Over the operating temperature range of the majority of electrical apparatus, the resistance of electrical conductors varies *linearly* with temperature, as shown in figure 1.2. The resistance of the conductor at $0°C$ is R_0, at some temperature θ_1 it is R_1, and at θ_2 it is R_2. From the figure we see that the slope of the graph is $(R_2 - R_1)/(\theta_2 - \theta_1)$, and is the increase in resistance per unit temperature rise. The increase in resistance expressed as a fraction of the resistance at temperature θ_1 is called the *temperature coefficient of resistance referred to* θ_1, and has the symbol α_1 where

$$\alpha_1 = \frac{R_2 - R_1}{(\theta_2 - \theta_1)R_1}$$

The temperature coefficient α_2, referred to θ_2 is

$$\alpha_2 = \frac{R_2 - R_1}{(\theta_2 - \theta_1)R_2}$$

In the case of metals, R_2 is always greater than R_1, and we say that the temperature coefficient of resistance has a **positive** value. In insulators, electrolytes,

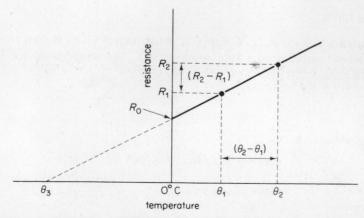

Figure 1.2   Variation of resistance with temperature

and many semiconductors, the value of R_2 is less than that of R_1, so that the graph slopes from left to right and the temperature coefficient has a **negative** value. In figure 1.2, the temperature coefficient of resistance referred to $0°C$ is α_0, where

$$\alpha_0 = \frac{R_1 - R_0}{\theta_1 R_0}$$

Rearranging this equation yields

$$R_1 = R_0(1 + \alpha_0 \theta_1) \tag{1.15}$$

Similarly

$$R_2 = R_0(1 + \alpha_0 \theta_2) \tag{1.16}$$

Hence

$$\frac{R_1}{R_2} = \frac{1 + \alpha_0 \theta_1}{1 + \alpha_0 \theta_2} \tag{1.17}$$

The value of α_0 can be computed from a knowledge of the value of θ_3, at which point the graph cuts the base line, when the resistance is zero.

$$\alpha_0 = \frac{R_0 - 0}{(0 - \theta_3)R_0} = \frac{1}{-\theta_3}$$

Measurements with annealed copper wire show that $\theta_3 = -234.5\,°C$, hence for copper

$$\alpha_0 = \frac{1}{-(-234.5)} = 0.004\,264 \quad (°C)^{-1}$$

Example 1.9

The resistance of a coil at $20\,°C$ was $20\,\Omega$. What current would it draw from a 10 V supply when operating in a cold room at $0\,°C$, the temperature coefficient of resistance being 0.0043 per $°C$ referred to $0\,°C$?

Solution

From equation 1.15

$$R_1 = R_0(1 + \alpha_0 \theta_1)$$

hence

$$R_0 = \frac{R_1}{1 + \alpha_0 \theta_1} = \frac{20}{1 + (0.0043 \times 20)} = 18.42\,\Omega$$

therefore

$$I_0 = \frac{E}{R_0} = \frac{10}{18.42} = 0.543 \text{ A}$$

Example 1.10

The field coil of an electric motor has a resistance of 250 Ω at 15 °C. By what amount, in Ω, will the resistance of the coil increase at the working temperature of 45°C? Assume $\alpha_0 = 0.0043$ per °C.

Solution

Using equation 1.17

$$\frac{R_1}{R_2} = \frac{1 + \alpha_0 \theta_1}{1 + \alpha_0 \theta_2}$$

hence

$$R_2 = \frac{R_1(1 + \alpha_0 \theta_2)}{(1 + \alpha_0 \theta_1)} = \frac{250[1 + (0.0043 \times 45)]}{1 + (0.0043 \times 15)}$$

$$= 280.3 \ \Omega$$

Resistance increase $= 280.3 - 250 = 30.3 \ \Omega$

1.12 Circuit Notation for Voltages and Currents

When writing down equations which define the operation of electrical circuits, it is convenient to use a notation that is easy to understand. A notation in common use is shown in figure 1.3, in which the current and voltages are indicated by arrows as follows.

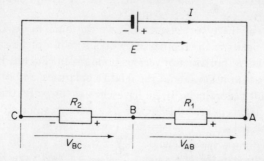

Figure 1.3   Electrical circuit voltage and current notations

(1) The *direction* of the current flowing in a *branch* or *arm* of a circuit is indicated by an arrow *on* the branch.

(2) The potential difference between individual points in the circuit is indicated by an *arrow between the points which is off the circuit*. The arrowhead on the *potential arrow* points towards the point which is *assumed* to be more positive.

An exception to the above notation occurs in one form of circuit analysis known as Maxwell's circulating-current method, discussed in chapter 2.

In figure 1.3, current flows out of the positive plate of the cell and enters R_1 via A, and leaves via B. In this case there is no doubt that point A is positive with respect to point B, so that we show the potential arrow pointing from B to A. Similarly, the potential arrow associated with R_2 points from C towards B. In cases of this kind we use a double subscript notation to define potential drops and, in the case of the p.d. across R_1, we say that the voltage across it is V_{AB}, where

$$V_{AB} = \text{the voltage at A relative to the voltage at B}$$

$$= (\text{the voltage at A}) - (\text{the voltage at B})$$

$$= V_A - V_B$$

Similarly

$$V_{BC} = \text{the voltage at B relative to the voltage at C}$$

$$= V_B - V_C$$

Since the cell is connected between A and C and, quite clearly, A is positive with respect to C, then

$$V_{AC} = E$$

also

$$V_{AC} = V_{AB} + V_{BC}$$

In a simple circuit of the type in figure 1.3, we can quickly decide which of two points is the more positive. In complex networks it may not be so easy to decide correctly on the relative polarities of various points in the circuit. In such a case we must simply make assumptions about the relative polarities, and draw the potential arrows on the circuit accordingly. If we have chosen correctly, then the calculated potential between those points will appear as a *positive* value. If we have chosen incorrectly, then the answer will be *negative*. For example if, in a complex network, we find that voltage V_{XY} has a value of +4 V t en point X is positive with respect to point Y by 4 volts. On the other hand, if $V_{XY} = -4$ V, then X is negative with respect to Y by 4 volts.

1.13 Resistors in Series and in Parallel

Circuit elements are said to be connected in series when each element carries the same current as the others. The three resistors in figure 1.4a are, by the above definition, in series with one another. The potential drop in R_1 is IR_1, that in R_2 is IR_2, and that in R_3 is IR_3. From the work in section 1.12, we see that the sum of the potential drops in the three resistors is equal to E, hence

$$E = IR_1 + IR_2 + IR_3 = I(R_1 + R_2 + R_3) = IR_S$$

where R_S is the **equivalent resistance** of the three resistors in series. Thus the circuit in figure 1.4b is electrically equivalent to figure 1.4a, in which

$$\left[R_S = R_1 + R_2 + R_3 \right.$$

In a circuit containing n resistors R_1, R_2, ..., R_n in series, the equivalent resistance of the circuit is

$$R_S = R_1 + R_2 + \ldots + R_n$$

The series connection results in an equivalent series resistance which is **greater** than the value of the highest individual value of resistance in the circuit.

The **equivalent conductance** of a series circuit is evaluated as follows. The p.d. across R_1 is I/G_1, where $G_1 = 1/R_1$, the p.d. across R_2 is I/G_2, etc., and

$$E = I\left(\frac{1}{G_1} + \frac{1}{G_2} + \frac{1}{G_3} \right)$$

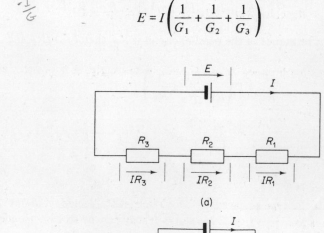

(a)

(b)

Figure 1.4   Resistors in series

Hence the equivalent conductance G_S of the series circuit is

$$G_S = \frac{I}{E} = \frac{1}{\dfrac{1}{G_1} + \dfrac{1}{G_2} + \dfrac{1}{G_3}}$$

The special case of two conductances in series (see, for example, section 2.9) yields the result

$$\left[G_S = \frac{1}{\dfrac{1}{G_1} + \dfrac{1}{G_2}} = \frac{G_1 G_2}{G_1 + G_2} \right]$$

Circuit elements are said to be in parallel with one another when the same voltage appears across each of the elements. The circuit in figure 1.5a shows three resistors in parallel, and it is shown below that the circuit may be replaced by the electrically equivalent circuit in figure 1.5b.

Since the sum of the currents flowing into the three parallel branches must be equal to the current flowing from the cell, then

$$I = I_1 + I_2 + I_3 = \frac{V}{R_1} + \frac{V}{R_2} + \frac{V}{R_3} = V\left(\frac{1}{R_1} + \frac{1}{R_2} + \frac{1}{R_3} \right)$$

Now, if the equivalent resistance of the parallel circuit is R_P, then $I = V/R_P$ and

$$\frac{V}{R_P} = V\left(\frac{1}{R_1} + \frac{1}{R_2} + \frac{1}{R_3} \right)$$

or

$$\frac{1}{R_P} = \frac{1}{R_1} + \frac{1}{R_2} + \frac{1}{R_3}$$

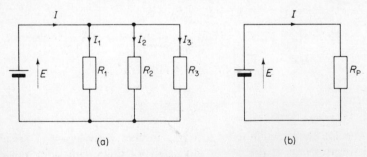

(a)   (b)

Figure 1.5   Resistors in parallel

The case of two resistors in parallel occurs in many circuits, and deserves special mention. Here the equivalent resistance of the circuit is

$$\left(\frac{1}{R_P} = \frac{1}{R_1} + \frac{1}{R_2} = \frac{R_1 + R_2}{R_1 R_2} \right) \qquad R_P = \frac{R_1 R_2}{R_1 + R_2}$$

or

$$\left(R_P = \frac{R_1 R_2}{R_1 + R_2} \right).$$

$$= \frac{\text{product of the two resistances}}{\text{sum of the two resistances}}$$

In a circuit containing n resistors in parallel, the equivalent resistance is

$$R_P = 1 \Big/ \left(\frac{1}{R_1} + \frac{1}{R_2} + \ldots + \frac{1}{R_n} \right)$$

Alternatively, the **equivalent conductance** G_P of the parallel circuit is

$$\left[G_P = G_1 + G_2 + G_3 \right]$$

where $G_P = 1/R_P$, $G_1 = 1/R_1$, $G_2 = 1/R_2$, etc.

Example 1.11

A circuit containing three series-connected resistances has an effective resistance of 100 Ω. If two of the resistors have values of 25 Ω and 60 Ω, respectively, what is the value of the third resistor?

Solution

$$R_S = R_1 + R_2 + R_3$$

where $R_S = 100 \,\Omega$, $R_1 = 25 \,\Omega$, $R_2 = 60 \,\Omega$, and the value of R_3 is unknown.

$$R_3 = 100 - (25 + 60) = 15 \,\Omega$$

Example 1.12

Four resistors are connected in series with one another, three having values of 5, 10, and 15 Ω, respectively. If the power dissipated by the circuit is 50 W when it is connected to a 50 V supply, what is the value of the unknown resistance?

Solution $p \propto VI$ or $I = \frac{V}{R}$ $\therefore P = \frac{V^2}{R}$

$$P = \frac{V^2}{R_S}$$

hence

$$R_S = \frac{V^2}{P} = \frac{50^2}{50} = 50 \ \Omega$$

Now

$$R_S = R_1 + R_2 + R_3 + R_4$$

Therefore

$$R_4 = 50 - (5 + 10 + 15) = 20 \ \Omega$$

Example 1.13

A parallel circuit contains resistors of 5, 10, and 15 Ω. Calculate the equivalent resistance of the parallel combination.

Solution

$$\frac{1}{R_P} = \frac{1}{5} + \frac{1}{10} + \frac{1}{15} = 0.2 + 0.1 + 0.0667 = 0.3667 \ \text{S}$$

Hence

$$R_P = 1/0.3667 = 2.727 \ \Omega$$

1.14 Series–Parallel Circuits

Series–parallel circuits contain a combination of series and parallel sections, one example being shown in figure 1.6a. In order to evaluate the equivalent electrical resistance of the complete network, the equivalent resistance of each section is first determined from which the total resistance is calculated.

In figure 1.6b, R_{P1} is the parallel combination of R_1, R_2, and R_3, while R_{P2} is the parallel combination of R_5 and R_6. Resistors R_{P1}, R_4, and R_{P2} are then combined to give an equivalent series resistance R. The latter value is taken in combination with R_7 to give the equivalent resistance R_E of the complete network, see figure 1.6c.

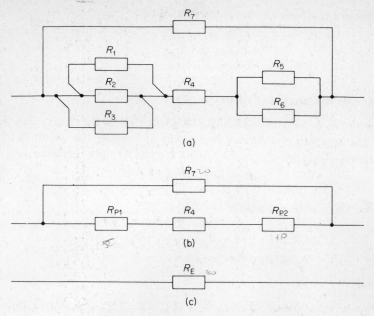

Figure 1.6   A series–parallel circuit

Example 1.14

In a circuit of the type in figure 1.6a, $R_1 = 10\ \Omega$, $R_2 = 20\ \Omega$, $R_3 = 20\ \Omega$, $R_4 = 5\ \Omega$, $R_5 = 30\ \Omega$, $R_6 = 15\ \Omega$, and $R_7 = 20\ \Omega$. Determine the equivalent resistance R_E of the complete circuit.

Solution

Adopting the procedure outlined above

$$\frac{1}{R_{P1}} = \frac{1}{R_1} + \frac{1}{R_2} + \frac{1}{R_3} = \frac{1}{10} + \frac{1}{20} + \frac{1}{20} = 0.2\ \text{S}$$

hence

$$R_{P1} = 1/0.2 = 5\ \Omega$$

and

$$R_{P2} = \frac{R_5 R_6}{R_5 + R_6} = \frac{30 \times 15}{30 + 15} = 10\ \Omega$$

therefore

$$R = R_{P1} + R_4 + R_{P2} = 5 + 5 + 10 = 20\ \Omega$$

The equivalent resistance of the complete circuit is

$$R_E = \frac{R_7 R}{R_7 + R} = \frac{20 \times 20}{20 + 20} = 10\ \Omega$$

1.15 Division of Current in Parallel Circuits *Current divider theorem.*

In many instances we need to calculate the magnitude of the current flowing in any one of the branches of a parallel circuit. Let us consider the circuit in figure 1.7. The current flowing into the parallel circuit is given by the expression

$$I = E/R_P$$

where $R_P = R_1 R_2 / (R_1 + R_2)$, and $I_1 = E/R_1$, hence

$$\frac{I_1}{I} = \frac{E/R_1}{E/R_P} = \frac{R_P}{R_1} = \frac{R_1 R_2 / (R_1 + R_2)}{R_1} = \frac{R_2}{R_1 + R_2}$$

therefore

$$\left[I_1 = \frac{R_2 I}{R_1 + R_2} \right]$$

It may also be shown that   *Current in branches.*

$$\left[I_2 = \frac{R_1 I}{R_1 + R_2} \right]$$

If we use the conductance of the paths in the parallel circuit, the following relationships are obtained

$$I_1 = \frac{G_1 I}{G_1 + G_2} \quad \text{and} \quad I_2 = \frac{G_2 I}{G_1 + G_2}$$

where $G_1 = 1/R_1$, and $G_2 = 1/R_2$.

In the general case where there are p parallel paths, we can show that the current

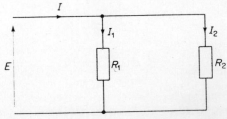

Figure 1.7   Division of current in a parallel circuit

flowing in the *m*th path is

$$\left[I_m = \frac{R_P}{R_m} I = \frac{G_m}{G_P} I \right]$$

where R_m and G_m are the resistance and conductance, respectively, of the *m*th path, and R_P and G_P are the equivalent resistance and conductance, respectively, of the parallel circuit.

Example 1.15

A parallel circuit containing two resistors of resistance 5 and 15 Ω, respectively, draws a current of 4 A from a power supply. Determine the current flowing in each resistor, the power consumed by each resistor, the total power consumed by the circuit, and the p.d. across the circuit.

Solution

Let $R_1 = 5\ \Omega$ and $R_2 = 15\Omega$.

$$I_1 = \frac{R_2 I}{R_1 + R_2} = \frac{15 \times 4}{5 + 15} = 3\ \text{A}$$

$$I_2 = \frac{R_1 I}{R_1 + R_2} = \frac{5 \times 4}{5 + 15} = 1\ \text{A}$$

The power consumed in R_1 is

$$P_1 = I_1^2 R_1 = 3^2 \times 5 = 45\ \text{W}$$

and in R_2 is

$$P_2 = I_2^2 R_2 = 1^2 \times 15 = 15\ \text{W}$$

The total power consumed by the circuit is the sum of the two values calculated above.

$$P = P_1 + P_2 = 45 + 15 = 60\ \text{W}$$

The p.d. across the circuit is equal to the p.d. across either of the resistors. For R_1 this is

$$E = I_1 R_1 = 3 \times 5 = 15\ \text{V}$$

1.16 Mechanical Quantities

In electrical engineering we need to use certain mechanical quantities, the units of these quantities being defined in terms of the units of length, mass, and time, namely the *metre*, the *kilogram*, and the *second* as follows.

Force

Symbol F. The **newton** (N) is that force which gives a mass of 1 kg an acceleration of 1 m/s^2. The relationship between these factors is

$$F = mf \quad \text{newtons}$$

where m = mass, and f = acceleration.

Energy or *Work*

Symbol W. The **joule** (J) is the work done when a force of 1 N acts through a distance of 1 m in the direction of the force, and

$$W = Fd \quad \text{joules}$$

where d = distance.

Thermal energy

Symbol Q. The energy gained or lost by a mass of m kilograms of substance when its temperature is changed by $\delta\theta$ kelvins (K) is

$$Q = m \times c \times \delta\theta \quad \text{joules}$$

where c is the *specific heat capacity* of the substance, whose dimensions are given as J/kg K or as kJ/kg K. Typical values of c are listed below.

Substance	*Value of c* (kJ/kg K)
Water	4.187
Iron	0.4187
Aluminium	0.385
Lead	0.1256

Torque

Symbol T, is the *turning moment* produced by a force about an axis or centre of rotation, and is the product of the force F which is at right angles to the radius of rotation R, where F is in newtons and R is in metres.

$$T = FR \quad \text{N m}$$

Velocity

This is the rate at which a body or particle is moving. *Linear velocity,* symbol v, has dimensions of m/s, and *angular velocity*, ω, has dimensions of rad/s. The two are related as follows

$$v = \omega R \quad \text{m/s}$$

where R = radius of rotation.

Power

Symbol P, is the rate of doing work, and the unit of power is the **watt (W)**, which is the joule/second.

$$P = \frac{Fd}{t} \quad \text{W or J/s}$$

$$= \text{force} \times \text{velocity} = Fv$$

The equation for *rotational power* produced by torque T is deduced below.

$$\text{rotational power} = \text{force} \times \text{velocity}$$

$$= \frac{T}{R} \times \omega R = \omega T \quad \text{W}$$

$$= \frac{2\pi NT}{60}$$

where N is the speed of revolution in revolutions per minute.

Example 1.16

The moving section of a linear motor has a mass of 2 kg, the frictional resistance to motion being negligible. If the field system causes the mass to accelerate at 200 m/s^2, what mechanical force is applied to the moving system?

Solution

$$F = mf = 2 \times 200 = 400 \text{ N}$$

Example 1.17

An electrical water-heater contains 22 kg of water at 15 °C. If the efficiency of the heater is 80 per cent, calculate the energy consumed by the heating element in order to raise the water temperature to 100 °C (a) in joules, and (b) in kWh. The specific heat capacity of water is 4.187 kJ/kg K.

Solution

The heat output from the element is

$$Q = m \times c \times \delta\theta$$

$$= 22 \times 4.187 \times (100 - 15) = 7830 \text{ kJ}$$

The equivalent heat input to the element is

$$\frac{7830}{0.8} = 9787 \text{ kJ}$$

and the electrical input in kWh is

$$\frac{9787 \times 1000}{3.6 \times 10^6} = 2.72 \text{ kWh}$$

Example 1.18

Determine the torque developed at the shaft of an electrical motor which provides an output power of 20 kW at a speed of 3000 revolutions per minute.

Solution

$$P = T\omega$$

hence

$$T = P/\omega \quad \text{N m}$$

The shaft speed in rad/s is given by the relationship

$$\omega = \frac{2\pi N}{60} = \frac{2\pi \times 3000}{60} = 314.2 \quad \text{rad/s}$$

hence

$$T = \frac{20\,000}{314.2} = 63.65 \quad \text{N m}$$

Summary of essential formulae and data

Electron: mass = 9.11×10^{-31} kilograms
              charge = -0.16×10^{-18} coulombs

Proton: mass = 1.67×10^{-27} kilograms
            charge = 0.16×10^{-18} coulombs

Charge: $Q = It$ coulombs

Ohm's law: $E = IR = I/G$ volts

Power: $P = EI = I^2 R = E^2/R$ watts

Energy: $W = EIt = I^2 Rt = E^2 t/R$ joules or watt seconds

Conductance: $G = 1/R$ siemens

Resistance: $R = \rho l/a = l/(\sigma a)$ ohms

Temperature coefficient: $\alpha_1 = \dfrac{R_2 - R_1}{(\theta_2 - \theta_1)R_1}$

Resistors in series: $R = R_1 + R_2 + \ldots$ ohms

Resistors in parallel: $1/R = 1/R_1 + 1/R_2 + \ldots$ ohms
$$G = G_1 + G_2 + \ldots \text{ siemens}$$

Division of current between R_1 and R_2 in parallel:
$$\text{current in } R_1 = \text{total current} \times R_2/(R_1 + R_2) \text{ amperes}$$

Force: $F = mf$ newtons

Energy: $W = Fd$ joules

Thermal energy: $Q = mc\delta\theta$ joules

Torque: $T = FR$ newton metres

Power: $P = Fd/t = \omega T$ watts

PROBLEMS

1.1 A 40-Ω resistor is connected to the terminals of an 80-V d.c. supply. Calculate (a) the current in the circuit, (b) the power consumed by the resistor, (c) the quantity of electricity passing through the circuit in 1 min, (d) the energy (i) in joules (ii) in kWh consumed in 50 min.
[(a) 2A; (b) 160 W; (c) 120 C; (d) (i) 480 000 J, (ii) 0.133 kWh]

1.2 Tests on a 'linear' and on a 'non-linear' resistor gave the following results.

V (V)	0	50	100	150	200
I_L (mA, linear resistor)	0	25	50	75	100
I_N (mA, non-linear resistor)	0	25	100	225	400

(a) Calculate the resistance R of the linear resistor. (b) If the relationship between the voltage and the current in the non-linear resistor is of the form $I = aV^n$, where I is in A and V in volts, determine the value of a and n.
[(a) 2 kΩ; (b) 10^{-5}, 2]

1.3 Three resistors are colour-coded respectively (from the most significant figure) (a) grey, red, orange, gold, (b) brown, red, silver, silver, (c) brown, green, green, no band. What are the resistor values?
[(a) 82 kΩ ± 5%; (b) 0.12 Ω ± 10%; (c) 1.5 MΩ ± 20%]

1.4 Determine the resistance of 1.5 km of copper wire having a resistivity of 1.73×10^{-8} Ω m and a cross-sectional area of 10 mm^2.
[2.6 Ω]

1.5 A conductor consisting of 40 strands of aluminium wire each of 1 mm diameter, the strands being twisted together, has a length of 500 m. If the effect of twisting the strands together results in an average increase in length of 2.5 per cent, calculate the resistance of the conductor if the conductance of aluminium is 35.33 $\times 10^6$ $(\Omega$ m$)^{-1}$.
[0.462 Ω]

1.6 A heater coil has 500 turns of wire. If the resistivity of the wire at the working temperature is 1.1 $\mu\Omega$ m and the coil is wound on a former of diameter 25 mm, calculate the resistance of the wire at its working temperature if the cross-sectional area of the wire is 0.5 mm^2.
[86.4 Ω]

1.7 If the temperature coefficient of resistance of a conductor referred to 0 $^\circ$C is 0.0042 $(^\circ$C$)^{-1}$ and its resistance at 0 $^\circ$C is 100 Ω, calculate its resistance at 80 $^\circ$C.
[133.6 Ω]

1.8 If the temperature coefficient of resistance of a material referred to 20 $^\circ$C is 0.005 $(^\circ$C$)^{-1}$, what is its value referred to 0 $^\circ$C?
[0.0056 $(^\circ$C$)^{-1}$]

1.9 A d.c. voltage of 250 V is applied to a coil of wire at a temperature of 15 $^\circ$C, and the current is 2.5 A. Some time later the temperature of the coil reached 25 $^\circ$C, at which time the current had fallen to 2.4 A. Calculate the temperature coefficient of the wire referred to 0 $^\circ$C.
[0.0044 $(^\circ$C$)^{-1}$]

1.10 Four similar electrical resistors are connected in parallel with each other. If their effective resistance is 92 Ω, determine the resistance of each resistor.
[23 Ω]

1.11 A resistance of 8 Ω is connected in series with another resistance of 4 Ω. What resistance must be connected in series with them to give a combined conductance of 0.05 S?
[8 Ω]

1.12 Determine the value of resistance R in figure 1.8 to give a total resistance of 1.0 Ω between terminals A and B.
[0.5 Ω]

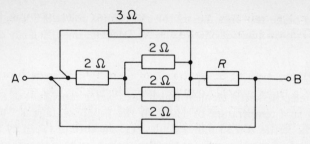

Figure 1.8

1.13 Calculate the effective resistance between terminals C and D in figure 1.9. If the circuit is connected to a 12-V d.c. supply, calculate the current drawn by the circuit and the power consumed by it.

[1.2 Ω; 10 A; 120 W]

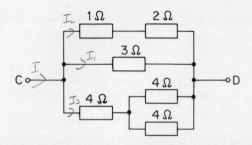

Figure 1.9

1.14 In the circuit in figure 1.9, calculate the current in the 2-Ω resistor and the power consumed by the 3-Ω resistor.

[4 A; 48 W]

1.15 Calculate the value of resistor *R* in figure 1.10 if the current in it is 8 A. Determine also the voltage across the parallel-connected section of the circuit and the power consumed by the 50-Ω resistor.

[12.5 Ω; 100 V; 200 W]

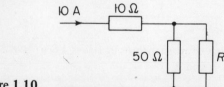

Figure 1.10

1.16   During a brake test on a d.c. motor, the output power is 8 kW and the input current at the rated voltage of 350 V is 26 A. Calculate the efficiency of the motor at this load.
[0.879 p.u.]

1.17   A 2-kW electric kettle element which has a water equivalent of 80 g contains 2 kg of water at a temperature of 16°C. If the supply voltage is 240 V and the efficiency of the kettle is 0.85 p.u., calculate (a) the current taken by the element, (b) the resistance of the element, (c) the time taken to heat the water to 100°C. Assume that the specific heat capacity of water is 4.187 kJ/kg K.
[(a) 8.33 A; (b) 28.8 Ω; (c) 7 m 10.3 s]

2 Circuit Theorems

The circuit theorems outlined in this chapter are illustrated by applications to d.c. circuits, but the theorems themselves can also be applied to a.c. circuits.

2.1 Kirchhoff's Laws

First law: The total current flowing towards a junction or node in a circuit is equal to the total current flowing away from the node, that is, the algebraic sum of the currents flowing towards the node is zero.

This law expressed mathematically is

$$\Sigma I = 0 \quad \text{at each node}$$

where the symbol Σ means 'the algebraic sum of'. In figure 2.1a, the current flowing towards node N is $(I_1 + I_3)$, and that flowing away from it is $(I_2 + I_4)$, hence

$$I_1 + I_3 = I_2 + I_4$$

or

$$(I_1 + I_3) - (I_2 + I_4) = 0$$

that is

$$\Sigma I = 0 \quad \text{at node N}$$

Second law: In any closed circuit, the algebraic sum of the potential drops and e.m.f.s in zero.;

This is expressed in mathematical form as

$$\Sigma IR + \Sigma E = 0$$

This statement is best understood by reference to figure 2.1b. The circuit contains two e.m.f.s and, in this case, we arbitrarily decided that E_1 is greater than E_2, so that the circuit current circulates around the loop in a clockwise direction. Having made this decision, we then draw a 'potential drop' arrow by the side of each resistor in the manner outlined in section 1.12. Starting at any point in the circuit, say point A, we proceed round the complete circuit writing down the values of the e.m.f.s

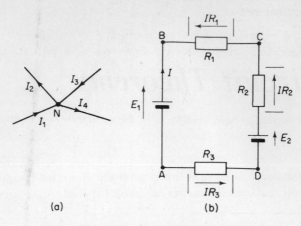

(a)   (b)

Figure 2.1   Illustrating Kirchhoff's laws

and p.d.s as we do so, giving each e.m.f. or p.d. a positive algebraic sign if the 'potential' arrow points in the direction in which we are moving around the circuit, and giving it a negative algebraic sign if the arrow points in the opposite directio. Suppose we elect to move round the loop via the sequence ABCDA (*note we must* always return to the starting point), then the loop equation is

$$E_1 - IR_1 - IR_2 - E_2 - IR_3 = 0$$

or

$$E_1 - E_2 = I(R_1 + R_2 + R_3) \tag{2.1}$$

that is

$$\Sigma E = \Sigma IR$$

Alternatively, if we proceed in the opposite direction, that is, in the direction ADCBA, the loop equation is

$$IR_3 + E_2 + IR_2 + IR_1 - E_1 = 0$$

or

$$I(R_1 + R_2 + R_3) = E_1 - E_2 \tag{2.2}$$

Since equations 2.1 and 2.2 are identical to one another, then we see that the *direction* in which we proceed around the network has no effect on the final equation.

There are two basic methods of applying Kirchhoff's laws, namely the *branch-current method* and *Maxwell's circulating-current method* (a method first suggested by James Clerk Maxwell, the Scottish physicist).

The branch-current method

In this method of analysis, a current is *assigned to each branch* of the network, after which Kirchhoff's laws are applied, as illustrated in the following examples.

Example 2.1

Calculate the current in each branch of the network shown in figure 2.2.

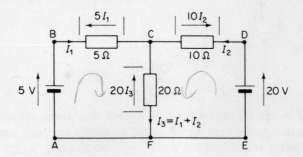

Figure 2.2

Solution

First, we assign currents I_1, I_2, and I_3 to the three branches joining at node C, noting that $I_3 = I_1 + I_2$. Next, we assign a 'potential' arrow to each e.m.f. and resistor, the direction in which the arrow points indicating which end of the resistor has the more positive potential *if* the current flows in the direction we have selected. The circuit equations can now be formed.

Loop ABCFA

$$5 - 5I_1 - 20(I_1 + I_2) = 0$$

or

$$5 = 25I_1 + 20I_2 \qquad (2.3)$$

Loop ABCDEFA

$$5 - 5I_1 + 10I_2 - 20 = 0$$

or

$$-15 = 5I_1 - 10I_2 \qquad (2.4)$$

We solve between equations 2.3 and 2.4 for currents I_1 and I_2. Multiplying equation 2.4 by 2 and adding it to equation 2.3 yields

$$-25 = 35I_1$$

or

$$I_1 = -\frac{25}{35} = -0.71 \text{ A}$$

Substituting this value into equation 2.3 gives

$$5 = [25 \times (-0.71)] + 20\,I_2$$

hence

$$I_2 = 1.14 \text{ A}$$

therefore

$$I_3 = I_1 + I_2 = (-0.71) + 1.14 = 0.43 \text{ A}$$

Since I_1 and I_3 have positive signs associated with them, it follows that these currents flow in the directions shown on the diagram. Current I_2 has a negative sign associated with it, so that it flows in the opposite direction to that chosen, that is, it flows from node C to node D.

Example 2.2

Determine the voltage V_{AB} in figure 2.3.

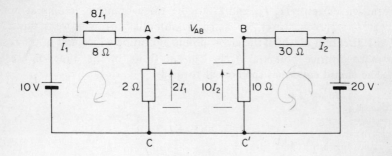

Figure 2.3

Solution

The circuit has two meshes which are connected at a·common point by the link C′C. No current flows between points A and B, but a potential exists between them. In order to determine the voltage V_{AB}, we need to evaluate the p.d.s in the path connecting the points A and B. We assign currents I_1 and I_2 to the loops in the manner shown, in which

$$I_1 = \frac{10}{8+2} = 1 \text{ A} \qquad I_2 = \frac{20}{10+30} = 0.5 \text{ A}$$

Now

$$V_{AB} = V_{C'B} + V_{CC'} + V_{AC}$$
$$= -10\,I_2 + 0 + 2\,I_1 \quad \text{using individual values of resistors between each letter.}$$
$$= -(10 \times 0.5) + (2 \times 1) = -3 \text{ V}$$

That is, the potential of point A with respect to point B is −3 V. Conversely, $V_{BA} = +3$ V.

Maxwell's circulating-current method

In this method of circuit analysis a *clockwise circulating-current is assigned to each mesh* in the network, the network equations being obtained by the application of Kirchhoff's laws. Certain of the branches will then carry two fictitious currents that flow in opposite directions, as occurs in the 20 Ω resistor in figure 2.4.

Example 2.3

Calculate the current in each branch of the network in figure 2.4.

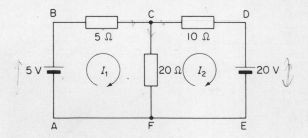

Figure 2.4

Solution

The circuit is seen to be identical to that in example 2.1. First, we assign currents I_1 and I_2 to the *meshes*, and note that both currents flow in the 20 Ω resistor, but in opposite directions to one another. The circuit equations are obtained by applying Kirchhoff's second law to each loop, restricting each path to the confines of one loop.

Loop ABCFA

$$5 - 5I_1 - 20I_1 + 20I_2 = 0$$

or

$$5 = 25I_1 - 20I_2 \qquad (2.5)$$

Loop CDEFC

$$-10I_2 - 20 - 20I_2 + 20I_1 = 0$$

or

$$-20 = 30I_2 - 20I_1 \qquad (2.6)$$

Solving between equations 2.5 and 2.6 yields

$$I_1 = -0.71 \text{ A}$$

$$I_2 = -1.14 \text{ A}$$

That is, both I_1 and I_2 flow in an anticlockwise direction in the actual circuit.

2.2 Thévenin's Theorem

Any two-terminal electrical network (sometimes known as a *one-port* network) can be replaced by an equivalent network which comprises an ideal voltage source in series with a resistance. Thus the circuit appearing between terminals A and B in figure 2.5a can be replaced by the equivalent circuit in figure 2.5b. This is summarised by **Thévenin's theorem** as follows.

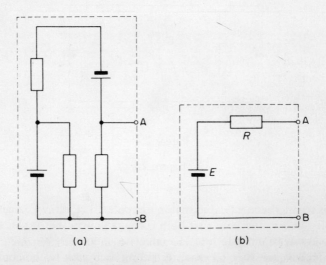

(a)   (b)

Figure 2.5   Illustrating the principle of Thévenin's theorem

An active network having two terminals A and B to which an electrical load may be connected, behaves as if the network contained a single source of e.m.f. E having an internal resistance R, where E is the p.d. measured between A and B with the load disconnected, and R is the resistance of the network between the points A and B when all the sources of e.m.f. within the network have been replaced by their internal resistances.

Resistance R is sometimes described as the *internal resistance* or *output resistance* of the equivalent network.

Example 2.4

Determine the current I_1 flowing in the 20 Ω resistor in figure 2.6a.

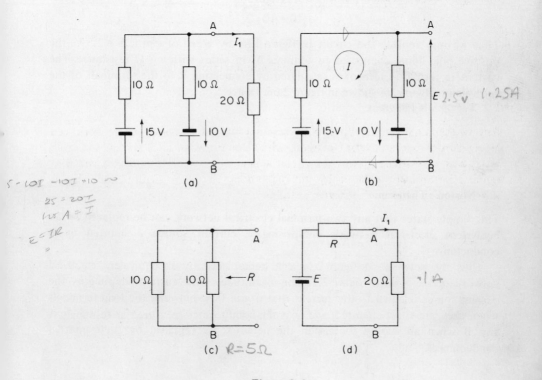

Figure 2.6

Solution

We first disconnect the 20 Ω resistor, as shown in figure 2.6b and calculate the voltage E appearing between terminals A and B. To do this, we assume that current I circulates between the two cells. Since the two e.m.f.s assist one another, then

$$I = \frac{(15 + 10)\text{V}}{(10 + 10)\Omega} = 1.25 \text{ A}$$

The p.d. between terminals A and B in figure 2.6b is

$$E = 15 - 10I = 15 - 12.5 = 2.5 \text{ V}$$

That is, the Thévenin equivalent circuit voltage source E has a value of 2.5 V, with point A being positive with respect to point B.

The value of R is determined by measuring the resistance between terminals A and B, having replaced the two cells meanwhile by their internal resistances (zero in this case). The resulting circuit is shown in figure 12.6c, giving

$$R = \frac{10 \times 10}{10 + 10} = 5 \ \Omega$$

Thus we may replace the circuit in figure 2.6b by a cell of e.m.f. $E = 2.5$ V, the positive pole being connected to terminal A, in series with a 5 Ω resistance. The current in the 20 Ω resistor is calculated by connecting it to the terminals of the equivalent network, as shown in figure 2.6d. Hence

$$I_1 = \frac{E}{R + 20} = \frac{2.5}{5 + 20} = 0.1 \text{ A}$$

$$h_1 = \frac{E}{I_1} = \frac{2.5}{5}$$

$$= .5 \ \Omega$$

2.3 Norton's Theorem

This simply states that **any two-terminal electrical network can be replaced by an equivalent electrical network comprising a current source I shunted by a conductance G.**

Thus the network in figure 2.5a can be replaced by an equivalent electrical network of the type in figure 2.7. The magnitude of the current supplied by the current source is equal to the current that would flow between the load terminals when they are short-circuited, and G is the conductance measured at terminals A and B when all supply sources in the network are replaced by their internal conductances.

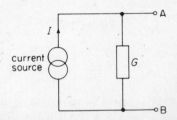

Figure 2.7 A current generator

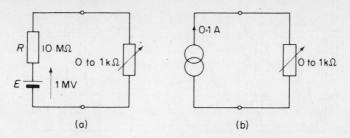

Figure 2.8 A physical interpretation of a current source

The concept of a current source is not an easy one to grasp, and it is best understood from the following analogy. A current source is a generator which provides a constant current into every value of load resistance. Ideally, the load resistance can vary from zero to infinity but, in practice, the maximum value of load resistance has a finite value. Suppose we wish to construct a current source that will provide a constant current of 0.1 A into a load whose resistance can have any value between zero and 1 kΩ. Such a source *could* consist of a voltage of 1 MV in series with a 10 MΩ resistor as shown in figure 2.8a; whatever value of load resistance in the range 0–1 kΩ were connected between the terminals, the load current would always be about 0.1 A. Thus the circuits in figures 2.8a and b may be regarded as equivalent to one another. From the above, we see that the output resistance of an ideal current source is infinity, that is, its output conductance is zero. In figure 2.8, any values of E and R could be used provided that the value of R is much greater than the maximum value of load resistance and also that $E/R = 0.1$.

Example 2.5

Using Norton's theorem, calculate the value of the current I_1 flowing in the 20 Ω resistor in figure 2.6a.

Solution

The magnitude of the equivalent current source is evaluated by applying a short-circuit to the load terminals and calculating the value of the short-circuit current, I, shown in figure 2.9a. Since the terminals are short-circuited

$$I_1 = \frac{15 \text{ V}}{10 \text{ }\Omega} = 1.5 \text{ A}$$

and

$$I_2 = \frac{10 \text{ V}}{10 \text{ }\Omega} = 1 \text{ A}$$

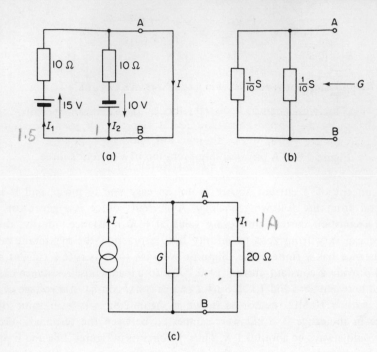

1.5

Figure 2.9

but

$$I_1 = I_2 + I$$

hence

$$I = I_1 - I_2 = 0.5 \text{ A} \qquad h_f$$

That is, the value of the Norton equivalent current generator is 0.5 A, with the current leaving terminal A and entering terminal B.

To determine the output conductance of the network with the 20 Ω load disconnected, the cells are replaced by their internal resistances (zero) – see figure 2.9b – and the conductance between the load terminals is calculated.

$$G = \frac{1}{10} + \frac{1}{10} = 0.2 \text{ S} \qquad h_o$$

The 20 Ω load is connected between the terminals of the Norton's equivalent circuit of the network in the manner shown in figure 2.9c, and the load current is calculated using the technique outlined in section 1.15 for current sharing in parallel circuits.

$$I_1 = I \times \frac{1/20}{G + 1/20} = 0.5 \times \frac{0.05}{0.2 + 0.05} = 0.1 \text{ A}$$

2.4 The Relationship between Thévenin's and Norton's Circuits

Since all active networks can be replaced either by the Thevenin equivalent circuit, figure 2.10a, or by the Norton equivalent circuit, figure 2.10b, then the parameters of the two circuits are related to one another. For Thevenin's circuit

$$V = E - I_1 R$$

hence

$$I_1 = \frac{E}{R} - \frac{V}{R}$$

or

$$\frac{E}{R} = I_1 + \frac{V}{R} \tag{2.7}$$

For Norton's circuit

$$I = I_1 + VG \tag{2.8}$$

For the two circuits to be equivalent to one another, then equations 2.7 and 2.8 must also be equivalent in every respect. That is

$$I = \frac{E}{R}$$

and

$$G = \frac{1}{R}$$

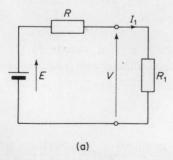 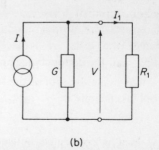

(a)   (b)

Figure 2.10   The relationship between (a) Thévenin's equivalent circuit and (b) Norton's equivalent circuit

The above relationships are verified by the results of examples 2.4 and 2.5. In example 2.4 we saw that the Thévenin equivalent circuit of figure 2.6a had a value for E of 2.5 V, and for R of 5 Ω. Applying the results above, we see that the Norton equivalent generator should have a current generator of $I = 2.5/5 = 0.5$ A, shunted by a conductance of $G = 1/5 = 0.2$ S. The latter values were obtained for figure 2.6a in example 2.5.

2.5 The Superposition Theorem

In a linear circuit containing several sources of e.m.f., the resultant current in any branch is the algebraic sum of the currents in that branch which would be produced by each e.m.f. acting alone, all other sources of e.m.f. being replaced meanwhile by their respective internal resistances.

This principle is not confined to electrical circuits, and may be applied to many forms of physical and mechanical systems.

The theorem is illustrated in its most basic form in example 2.6.

Example 2.6

Calculate the value of the current I in figure 2.11a using the superposition principle.

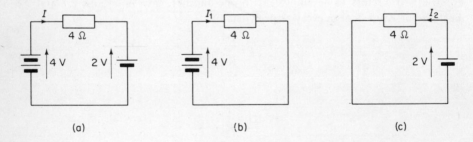

(a)   (b)   (c)

Figure 2.11

Solution

The current distribution due to each e.m.f. taken separately is calculated, the net circuit current being the sum of the two currents. In figure 2.11b, current I_1 due to the 4 V cell taken alone is

$$I_1 = 4/4 = 1 \text{ A}$$

and flows in a *clockwise* direction around the circuit. Current I_2 due to the 2 V cell taken alone is

$$I_2 = 2/4 = 0.5 \text{ A}$$

and flows in an *anticlockwise* direction. Applying the principle of superposition, the current *I* flowing in a *clockwise* direction in figure 2.11a is

$$I = I_1 - I_2 = 1 - 0.5 = 0.5 \text{ A}$$

Alternatively, we can say that a current of −0.5 A flows in an anticlockwise direction.

Example 2.7

Calculate the current flowing in each branch of the circuit in figure 2.12a.

Solution

Removing the 5 V cell and replacing it by its internal resistance (zero in this case) gives the circuit in b. The effective resistance connected to the 10 V battery terminals is

$$10 + \frac{30 \times 20}{30 + 20} = 22 \ \Omega$$

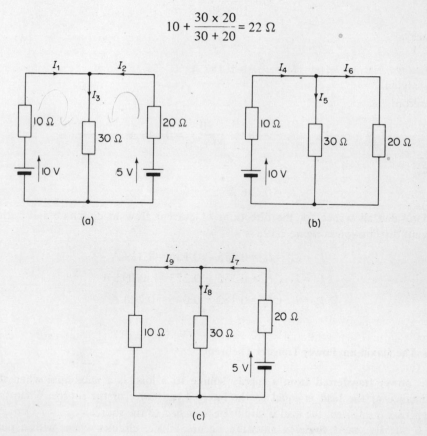

Figure 2.12

hence

$$I_4 = \frac{10}{22} = 0.455 \text{ A}$$

Using the rules for the division of current in a parallel circuit

$$I_5 = 0.455 \times \frac{20}{20 + 30} = 0.182 \text{ A}$$

$$I_6 = 0.455 \times \frac{30}{20 + 30} = 0.273 \text{ A}$$

Next, we remove the 10 V cell from figure 2.12a and replace it by its internal resistance (zero), and calculate the current distribution in the resulting network, figure 2.12c. The resistance presented to the terminals of the 5 V battery is

$$20 + \frac{10 \times 30}{10 + 30} = 27.5 \ \Omega$$

hence

$$I_7 = \frac{5}{27.5} = 0.182 \text{ A}$$

therefore

$$I_8 = 0.182 \times \frac{10}{10 + 30} = 0.046 \text{ A}$$

and

$$I_9 = 0.136 \text{ A}$$

Making due allowance for the directions of current flow in circuits b and c, the current distribution in figure 2.12a is

$$I_1 = I_4 - I_9 = 0.455 - 0.136 = 0.319 \text{ A}$$
$$I_2 = I_7 - I_6 = 0.182 - 0.273 = -0.091 \text{ A}$$
$$I_3 = I_5 + I_8 = 0.182 + 0.046 = 0.228 \text{ A}$$

2.6 The Maximum Power Transfer Theorem

The power transferred from a supply source to a load is a maximum when the resistance of the load is equal to the internal resistance of the source. When this condition is satisfied, the load is said to be *matched* to the source.

A slightly more complex situation occurs in a.c. circuits when, with a pure resistive load, maximum power is transferred into the load when the load resistance is equal to the modulus of the impedance of the supply source (impedance is dealt with in chapter 6).

The theorem is verified for the pure resistive case by reference to figure 2.13. Here

$$I = \frac{E}{r + R}$$

and the power consumed by the load is

$$P_L = I^2 R = \frac{E^2 R}{(r + R)^2} = \frac{E^2 R}{r^2 + 2Rr + R^2}$$

$$= \frac{E^2}{(r^2/R) + 2r + R} \tag{2.9}$$

Clearly, when $R = 0$ the power consumed by the load is zero. When $R = \infty$ the load power is also zero since $I = 0$. Between the two conditions lies a value of R for which the power consumed is a maximum. This occurs when the denominator of equation 2.9 is a minimum. To determine this condition, we differentiate the denominator of equation 2.9 with respect to R, and equate the result to zero.

$$\frac{\mathrm{d}}{\mathrm{d}R} \left[(r^2/R) + 2r + R \right] = - \left(\frac{r}{R} \right)^2 + 1$$

hence

$$- \left(\frac{r}{R} \right)^2 + 1 = 0$$

or

$$R = r \tag{2.10}$$

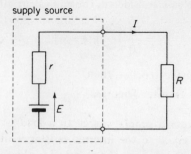

supply source

Figure 2.13 The maximum power transfer theorem

To verify that the condition in equation 2.10 does, in fact, give the denominator of equation 2.9 a minimum value, we must check that the differential of $[1 - (r/R)^2]$ with respect to R has a positive value.

$$\frac{d}{dR}\left[1 - \left(\frac{r}{R}\right)^2\right] = \frac{2r^2}{R^3}$$

which has a positive value.

Hence, from equation 2.10, for maximum power to be transferred into a resistive load, the load and source resistances must be equal in value.

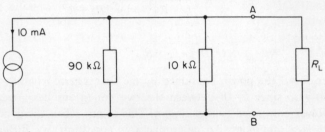

Figure 2.14

Example 2.8

The circuit in figure 2.14 is the equivalent output circuit of a transistor amplifier. Determine (a) the value of the load resistor R_L to give maximum power transfer, and (b) the magnitude of the power delivered into the load.

Solution

The output resistance of the network is obtained by measuring the resistance between points A and B, the current generator being meanwhile replaced by its internal resistance (infinity, since it is a current generator) and the load resistor being disconnected. In the case considered, the equivalent output resistance is

$$r = \frac{10 \times 90}{10 + 90}\,k\Omega = 9\,k\Omega$$

For maximum power transfer to occur, then

$$R_L = r = 9\,k\Omega$$

(b) The current in R_L is

$$I_L = 10 \times \frac{r}{R_L + r}\,mA = 10 \times \frac{9}{9 + 9} = 5\,mA$$

and the power consumed by R_L is

$$P_L = (5 \times 10^{-3})^2 \times 9000 = 0.225 \text{ W}$$

2.7 Compensation Theorem or Substitution Theorem

A network of resistance R which carries current I may be replaced by a *compensation e.m.f.* or *substitution e.m.f.* whose magnitude and polarity are equal to the p.d. IR. Also, if the voltage across an element or branch of resistance R is V, then the element or branch may be replaced by a current source of $I = V/R$.

This theorem can be extended to provide a method for estimating the change in current distribution in a system when the resistance of one of the elements or branches changes by an amount δR, that is, from R to $R \pm \delta R$, and is illustrated in example 2.10.

Example 2.9

In the network in figure 2.15a, replace R by a compensation e.m.f.

Solution

The effective resistance presented to the battery terminals is

$$100 + \frac{500 \times (50 + 150)}{500 + 50 + 150} = 242.9 \ \Omega$$

hence

$$I_1 = 100/242.9 = 0.412 \text{ A}$$

therefore

$$I = 0.412 \times \frac{500}{500 + 50 + 150} = 0.2943 \text{ A}$$

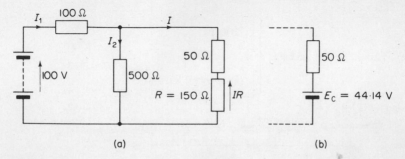

(a)   (b)

Figure 2.15

The magnitude of the compensation e.m.f. E_C is

$$E_C = IR = 0.294 \times 150 = 44.14 \text{ V}$$

The e.m.f. E_C is placed in the network in the manner shown in figure 2.15b.

Example 2.10

If, in figure 2.15a, the value of the 500-Ω resistor is *increased* in value to 500 Ω, calculate the new values of the branch currents.

Solution

To solve this type of problem we use both the compensation theorem and the superposition theorem. First, we calculate the original current distribution in the system. This was completed in example 2.9 to give the values

$$I_1 = 0.412 \text{ A} \quad \text{and} \quad I = 0.2943 \text{ A}$$

hence

$$I_2 = I_1 - I = 0.1177 \text{ A}$$

Next we re-draw the circuit diagram to show the circuit with the modified value of resistance of 550 Ω together with a compensation e.m.f. $I_2 \delta R$, where $\delta R = 50 \, \Omega$. In this circuit the original voltage source is replaced by its internal resistance (zero), the complete circuit to calculate the *changes in current distribution* being shown in figure 2.16. The changes in network current due to the introduction of the compensation e.m.f. are δI_1, δI_2, and δI. If the currents in the final circuit are I_1' I_2', and I', then bearing in mind the directions of flow of current, their values are

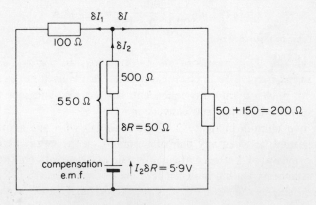

Figure 2.16

calculated from the relationships

$$I_1' = I_1 - \delta I_1$$
$$I_2' = I_2 - \delta I_2$$
$$I' = I + \delta I$$

The resistance presented to e.m.f. $I_2 \delta R$ is

$$550 + \frac{100 \times 200}{100 + 200} = 550 + 66.67 = 616.67 \ \Omega$$

The magnitude of the compensation e.m.f. is

$$I_2 \delta R = 0.1177 \times 50 = 5.9 \ \text{V}$$

hence

$$\delta I_2 = \frac{5.9}{616.67} = 0.0096 \ \text{A}$$

also

$$\delta I = \delta I_2 \times \frac{100}{100 + 200} = 0.0032 \ \text{A}$$

and

$$\delta I_1 = \delta I_2 \times \frac{200}{100 + 200} = 0.0064 \ \text{A}$$

The final values of current in the circuit are

$$I_1' = I_1 - \delta I_1 = 0.412 - 0.0064 = 0.4056 \ \text{A}$$
$$I_2' = I_2 - \delta I_2 = 0.1177 - 0.0096 = 0.1081 \ \text{A}$$
$$I' = I + \delta I = 0.2943 + 0.0032 = 0.2975 \ \text{A}$$

2.8 Delta–Star Transformation

The **delta** and **star** configurations of components, in figures 2.17a and b, respectively, occur frequently in electrical circuits. It is convenient in some cases to convert a delta network into its equivalent star circuit, and vice versa. In this section we deal with the delta–star conversion.

When the two circuits in figure 2.17 are identical with one another, then the resistance measured between any pair of terminals on the delta circuit is equal to the resistance measured between the same pair of terminals on the star circuit. Thus, the resistance appearing between terminals 1 and 2 on the delta circuit must have the same value as the resistance between terminals 1 and 2 on the star circuit,

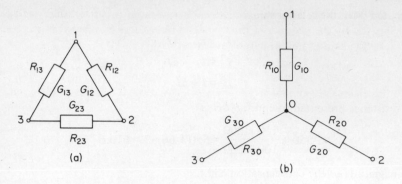

Figure 2.17   The delta–star transformation

that is

$$R_{10} + R_{20} = \frac{R_{12}(R_{13} + R_{23})}{R_{12} + R_{23} + R_{13}} \tag{2.11}$$

The resistance measured between terminals 2 and 3 is

$$R_{20} + R_{30} = \frac{R_{23}(R_{13} + R_{12})}{R_{12} + R_{23} + R_{13}} \tag{2.12}$$

and between terminals 1 and 3 is

$$R_{10} + R_{30} = \frac{R_{13}(R_{12} + R_{23})}{R_{12} + R_{23} + R_{13}} \tag{2.13}$$

Subtracting equation 2.12 from equation 2.11 gives

$$R_{10} - R_{30} = \frac{R_{12}R_{13} - R_{13}R_{23}}{R_{12} + R_{23} + R_{13}} \tag{2.14}$$

By adding equations 2.13 and 2.14 and dividing the result by 2, we obtain the following equation

$$R_{10} = \frac{R_{12}R_{13}}{R_{12} + R_{23} + R_{13}} \tag{2.15}$$

Using a similar technique, the following relationships are deduced

$$R_{20} = \frac{R_{12}R_{23}}{R_{12} + R_{23} + R_{13}} \tag{2.16}$$

$$R_{30} = \frac{R_{13}R_{23}}{R_{12} + R_{23} + R_{13}} \tag{2.17}$$

Hence, the equivalent resistance connected between an input terminal and the star point is given by the product of the two delta resistances connected to the terminal divided by the sum of the three resistances in the delta.

In some circuits the delta is shown in the form of a π-circuit — see figure 2.18a — and the star is shown as a T-circuit — see figure 2.18b. When converting from the π-circuit to the T-circuit, equations 2.15–2.17 hold good.

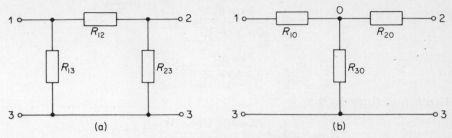

Figure 2.18   The π–T transformation

Example 2.11

Calculate the resistance between terminals A and B in figure 2.19a.

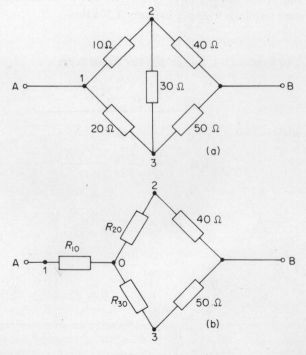

Figure 2.19

Solution

We first convert the set of delta-connected resistors between points 1, 2, and 3 in figure 2.19a into the star combination R_{10}, R_{20}, R_{30} in b as follows

$$R_{10} = \frac{10 \times 20}{10 + 20 + 30} = 3.33 \ \Omega$$

$$R_{20} = \frac{10 \times 30}{10 + 20 + 30} = 5 \ \Omega$$

$$R_{30} = \frac{20 \times 30}{10 + 20 + 30} = 10 \ \Omega$$

Hence, from figure 2.19b

$$R_{AB} = 3.33 + \frac{(5 + 40)(10 + 50)}{5 + 40 + 10 + 50} = 29.04 \ \Omega$$

2.9 Star–Delta Transformation

If we short-circuit terminals 2 and 3 in both the delta and star networks in figures 2.17a and b, the circuits in figures 2.20a and b remain. The *conductance* measured between terminals W and X in figure 2.20a is

$$G_{WX} = G_{12} + G_{13}$$

From the work in section 1.13, we see that for figure 2.20b

$$G_{YZ} = \frac{G_{10} (G_{20} + G_{30})}{G_{10} + G_{20} + G_{30}}$$

Equation the relationships for G_{WX} and G_{YZ} gives

$$G_{12} + G_{13} = \frac{G_{10} (G_{20} + G_{30})}{G_{10} + G_{20} + G_{30}} \tag{2.18}$$

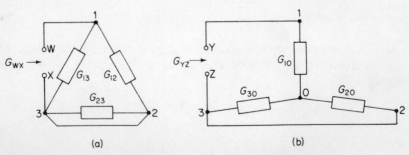

(a)   (b)

Figure 2.20   The star–delta transformation

If we now remove the above short-circuit and re-apply it between terminals 1 and 3 in both the star and delta networks, we obtain the relationship

$$G_{12} + G_{23} = \frac{G_{20}(G_{10} + G_{30})}{G_{10} + G_{20} + G_{30}} \qquad (2.19)$$

Repeating the above procedure between terminals 1 and 2 yields

$$G_{13} + G_{23} = \frac{G_{30}(G_{10} + G_{20})}{G_{10} + G_{20} + G_{30}} \qquad (2.20)$$

Manipulating equations 2.18–2.20 in much the same manner as we used equations 2.11–2.13 in section 2.8, the following results are obtained

$$G_{12} = \frac{G_{10}G_{20}}{G_{10} + G_{20} + G_{30}} \qquad (2.21)$$

$$G_{23} = \frac{G_{20}G_{30}}{G_{10} + G_{20} + G_{30}} \qquad (2.22)$$

$$G_{13} = \frac{G_{10}G_{30}}{G_{10} + G_{20} + G_{30}} \qquad (2.23)$$

The above expressions are related to the resistance values in the circuit as follows

$$R_{12} = \frac{1}{G_{12}} = R_{10} + R_{20} + \frac{R_{10}R_{20}}{R_{30}} \qquad (2.24)$$

$$R_{23} = \frac{1}{G_{23}} = R_{20} + R_{30} + \frac{R_{20}R_{30}}{R_{10}} \qquad (2.25)$$

$$R_{13} = \frac{1}{G_{13}} = R_{10} + R_{30} + \frac{R_{10}R_{30}}{R_{20}} \qquad (2.26)$$

Example 2.12

Calculate the resistance between points A and B in figure 2.21a.

Solution

We replace the star-connected combination of resistors of values 20, 30 and 50 Ω with a delta-connected set in the manner shown in figure 2.21b. In this circuit, from equation 2.24

$$R_{12} = R_{10} + R_{20} + \frac{R_{10}R_{20}}{R_{30}}$$

$$= 20 + 30 + \frac{20 \times 30}{50} = 62 \ \Omega$$

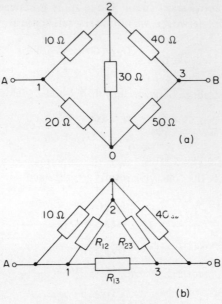

Figure 2.21

also

$$R_{23} = 30 + 50 + \frac{30 \times 50}{20} = 155 \ \Omega$$

and

$$R_{13} = 20 + 50 + \frac{20 \times 50}{30} = 103.33 \ \Omega$$

The parallel combination of R_{12} and $10 \ \Omega$ has a value of $8.61 \ \Omega$, and the parallel combination of R_{23} and $40 \ \Omega$ has a value of $31.79 \ \Omega$. Hence

$$R_{AB} = \frac{103.33 \ (8.61 + 31.79)}{8.61 + 31.79 + 103.33} = 29.04 \ \Omega$$

(See also the solution of example 2.11.)

2.10 Mesh or Loop Current Analysis

If we apply Maxwell's circulating-current method to a multi-mesh network, we find that the resulting equations follow a logical sequence, from which we can derive a method for writing down the equations by inspection. For the circuit in

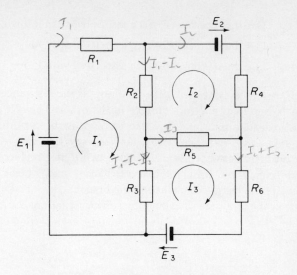

Figure 2.22   Mesh or loop analysis

figure 2.22, the mesh equations are

Mesh 1 $E_1 = I_1(R_1 + R_2 + R_3) - I_2 R_2 - I_3 R_3$

Mesh 2 $E_2 = -I_1 R_2 + I_2(R_2 + R_4 + R_5) - I_3 R_5$

Mesh 3 $E_3 = -I_1 R_3 - I_2 R_5 + I_3(R_3 + R_5 + R_6)$

The above equations can be re-written in a more convenient 'standard' form as follows

$$\left.\begin{aligned}
E_1 &= I_1 R_{11} + I_2 R_{12} + I_3 R_{13} \\
E_2 &= I_1 R_{21} + I_2 R_{22} + I_3 R_{23} \\
E_3 &= I_1 R_{31} + I_2 R_{32} + I_3 R_{33}
\end{aligned}\right\} \qquad (2.27)$$

where E_1 = total e.m.f. acting in loop 1; E_2 = total e.m.f. acting in loop 2; E_3 = total e.m.f. acting in loop 3.

Note The e.m.f.s in equations 2.27 all have positive signs since they act in a direction to produce a clockwise circulating current within their own mesh.

R_{11} = self-resistance of loop 1 = $R_1 + R_2 + R_3$
R_{22} = self-resistance of loop 2 = $R_2 + R_4 + R_5$
R_{33} = self-resistance of loop 3 = $R_3 + R_5 + R_6$

In order to determine the **self-resistance** of each loop, replace each generator in the loop by its internal resistance, and open-circuit *all* other loops, then calculate the loop resistance.

$R_{12} = R_{21} = (-1) \times$ the mutual resistance associated with loops 1 and 2 $= -R_2$

$R_{13} = R_{31} = (-1) \times$ the mutual resistance associated with loops 1 and 3 $= -R_3$

$R_{23} = R_{32} = (-1) \times$ the mutual resistance associated with loops 2 and 3 $= -R_5$

The mutual resistance associated with two loops is the resistance which causes a potential drop in one loop due to a current in the other. Thus R_{13} is the mutual resistance which causes a p.d. in loop 1 due to the flow of current I_3 in loop 3. If no such p.d. exists (see example 2.13), then we say that $R_{13} = 0$. In order to determine the **mutual resistance** associated with two loops, replace each generator in the mutual branch by their internal resistances and open-circuit *all* other branches. Calculate the resistance of the mutual branch.

The currents in the loops are calculated by solving equations 2.27.

Example 2.13

Derive the mesh equations for the circuit in figure 2.23.

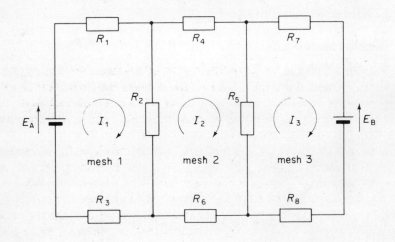

Figure 2.23

Solution

The self-resistance of loop 1 in which I_1 circulates is $(R_1 + R_2 + R_3)$, and the resistance in the mutual branch between meshes 1 and 2 is R_2, hence $R_{11} = R_1 + R_2 + R_3$ and $R_{12} = -R_2$. Since no branch exists that is common to both mesh 1 and mesh 3, then $R_{13} = 0$. Also, in mesh 1, e.m.f. E_A acts in a direction to produce a clockwise circulating current, then $E_1 = E_A$ and, for mesh 1

$$E_A = I_1(R_1 + R_2 + R_3) - I_2 R_2$$

In mesh 2

$$E_2 = 0$$
$$R_{21} = -R_2$$
$$R_{22} = R_2 + R_4 + R_5 + R_6$$
$$R_{23} = -R_5$$

hence, for mesh 2

$$0 = -I_2 R_2 + I_2(R_2 + R_4 + R_5 + R_6) - I_3 R_5$$

In mesh 3, e.m.f. E_B acts in a direction which **opposes** the clockwise circulation of current, so that $E_3 = -E_B$. In this mesh

$$R_{31} = 0$$
$$R_{32} = -R_5$$
$$R_{33} = R_5 + R_7 + R_8$$

therefore, for mesh 3

$$-E_B = -I_2 R_5 + I_3(R_5 + R_7 + R_8)$$

2.11 Nodal Voltage Analysis

Nodal analysis enables us to write down a set of equations for a given circuit in terms of the voltages appearing at each of the nodes in the circuit. Let us consider the application of nodal analysis to figure 2.24a. When using this technique, voltage sources are first converted into their equivalent current sources, and the resistance of each branch is converted into its equivalent conductance.

Thus the voltage source E_A together with its series resistance R_A is converted to its equivalent current source in figure 2.24b, and comprises current source $I_1 = E_A/R_A$ shunted by conductance $G_A = 1/R_A$ (see section 2.4). Also, voltage source E_B with its series resistance R_B is converted into current source $I_2 = E_B/R_B$ shunted by conductance $G_B = 1/R_B$. The conductances between pairs of nodes are then grouped together to simplify the circuit to the form shown in figure 2.24c, in which

$$G_1 = G_A + G_C = \frac{1}{R_A} + \frac{1}{R_C}$$

$$G_2 = G_B + G_E = \frac{1}{R_B} + \frac{1}{R_E}$$

Since all voltages in the circuit are measured with respect to node 3, we describe this node as the *reference node*, and all voltages are understood to be measured from this node. That is $V_1 = V_{13}$ and $V_2 = V_{23}$.

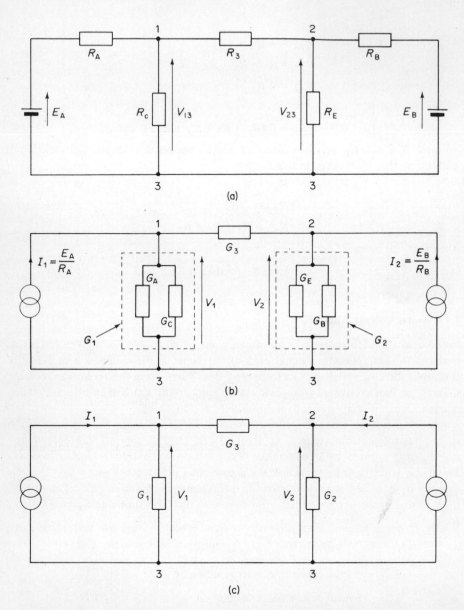

Figure 2.24   Nodal or voltage analysis

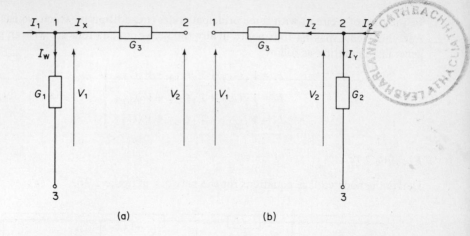

Figure 2.25 Conditions at (a) node 1, and (b) node 2

In order to write down the equations for the sum of the currents at each node, we isolate the nodes in the manner shown in figure 2.25. Since the total current flowing towards node 1 from the current source is I_1, then

$$I_1 = I_W + I_X = V_1 G_1 + (V_1 - V_2)G_3$$
$$= V_1 (G_1 + G_3) - V_2 G_3 \qquad (2.28)$$

For node 2

$$I_2 = I_Y + I_Z = V_2 G_2 + (V_2 - V_1)G_3$$
$$= - V_1 G_3 + V_2 (G_2 + G_3) \qquad (2.29)$$

Equations 2.28 and 2.29 are written in a more convenient generalised form below

$$\left. \begin{array}{l} I_1 = V_1 G_{11} + V_2 G_{12} \\ I_2 = V_1 G_{12} + V_2 V_{22} \end{array} \right\} \qquad (2.30)$$

where I_1 = the total current flowing *towards* node 1 from all current sources; I_2 = total current flowing *towards* node 2 from all current sources; and

$$G_{11} = \text{total conductance terminating on node 1} = G_1 + G_3$$

$$G_{22} = \text{total conductance terminating on node 2} = G_2 + G_3$$

also

$$G_{12} = (-1) \times \text{the conductance linking node 1 and node 2} = -G_3$$

$$G_{21} = (-1) \times \text{the conductance linking node 2 and node 1} = -G_3$$

In the case of a circuit with three principal nodes (in addition to which we have the reference node, making four nodes in all), the network equations are written in the generalised notation as follows

$$
\left.
\begin{aligned}
I_1 &= V_1 G_{11} + V_2 G_{12} + V_3 G_{13} \\
I_2 &= V_1 G_{21} + V_2 G_{22} + V_3 G_{23} \\
I_3 &= V_1 G_{31} + V_2 G_{32} + V_3 G_{33}
\end{aligned}
\right\}
\tag{2.31}
$$

Example 2.14

Derive the node voltage equations for the network in figure 2.26.

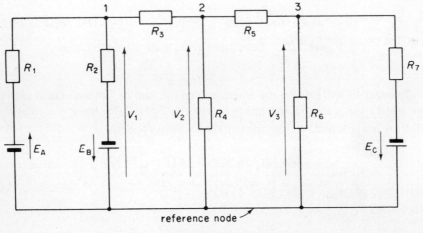

Figure 2.26

Solution

The three voltage generators are converted into their equivalent current generators and are connected into the circuit in the manner shown in figure 2.27, in which $I_A = E_A/R_1$, $I_B = E_B/R_2$, $I_C = E_C/R_7$, $G_1 = 1/R_1$, $G_2 = 1/R_2$, etc. The currents produced by these generators act in the same direction as the e.m.f.s of the voltage generators. Using the notation developed in this section

$$
\begin{aligned}
I_1 &= I_A - I_B \\
I_2 &= 0 \\
I_3 &= -I_C
\end{aligned}
$$

Readers will note that no current generators are connected to node 2, hence the current flowing into node 2 from current sources is zero, so that $I_2 = 0$. Also,

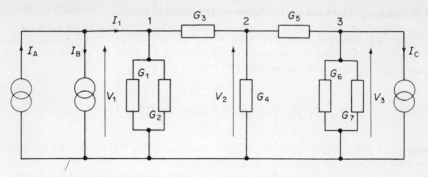

Figure 2.27 Example 2.14, showing the equivalent current generators

current I_C flows *away* from node 3, hence $I_3 = -I_C$. The total admittances terminating on the nodes are

$$G_{11} = G_1 + G_2 + G_3 = \frac{1}{R_1} + \frac{1}{R_2} + \frac{1}{R_3}$$

$$G_{22} = G_3 + G_4 + G_5 = \frac{1}{R_3} + \frac{1}{R_4} + \frac{1}{R_5}$$

$$G_{33} = G_5 + G_6 + G_7 = \frac{1}{R_5} + \frac{1}{R_6} + \frac{1}{R_7}$$

and the conductances linking the nodes are

$$G_{12} = G_{21} = -G_3 = -\frac{1}{R_3}$$

$$G_{23} = G_{32} = -G_5 = -\frac{1}{R_5}$$

$$G_{13} = G_{31} = 0$$

Since the circuit has three principal nodes, the nodal equations have the general form given in equation 2.31. Inserting the values above into equation 2.31 gives the following equations. Node 1

$$\frac{E_A}{R_1} - \frac{E_B}{R_2} = V_1\left(\frac{1}{R_1} + \frac{1}{R_2} + \frac{1}{R_3}\right) - \frac{V_2}{R_3}$$

$$0 = -\frac{V_1}{R_3} + V_2\left(\frac{1}{R_3} + \frac{1}{R_4} + \frac{1}{R_5}\right) - \frac{V_3}{R_5}$$

$$-\frac{E_C}{R_7} = -\frac{V_2}{R_5} + V_3\left(\frac{1}{R_5} + \frac{1}{R_6} + \frac{1}{R_7}\right)$$

2.12 Millman's Theorem or the Parallel-generator Theorem

The theorem outlined here represents a particular application of nodal analysis. This theory states that the voltage $V_{O'O}$ appearing between terminals O' and O in figure 2.28a is given by the expression

$$V_{O'O} = \frac{E_1 G_1 + E_2 G_2 + E_3 G_3}{\Sigma G}$$

where $G_1 = 1/R_1$, $G_2 = 1/R_2$, $G_3 = 1/R_3$, and $\Sigma G = G_1 + G_2 + G_3$. In the general case where there are n generators in parallel, then

$$V_{O'O} = \frac{\sum\limits_{k=1}^{k=n} E_{kO} G_k}{\sum\limits_{k=1}^{k=n} G_k}$$

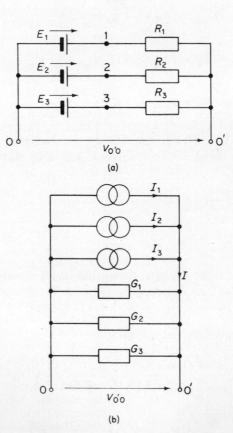

(a)

(b)

Figure 2.28   Millman's theorem

where E_{kO} is the potential of node k with respect to node O, and G_k is the conductance connecting node k to node O'.

To verify the theorem in the case of figure 2.28a, we convert the voltage sources into their equivalent current sources as shown in figure 2.28b, where $I_1 = E_1/R_1$, $I_2 = E_2/R_2$, and $I_3 = E_3/R_3$. The total current flowing towards node O' is $I = I_1 + I_2 + I_3$, and this current generates the voltage $V_{O'O}$ across conductance $(G_1 + G_2 + G_3)$, hence

$$I_1 + I_2 + I_3 = V_{O'O} (G_1 + G_2 + G_3)$$

or

$$V_{O'O} = \frac{E_1 G_1 + E_2 G_2 + E_3 G_3}{G_1 + G_2 + G_3}$$

Expressed in mathematical form, the above is given as

$$V_{O'O} = \frac{\sum_{k=1}^{k=3} E_{kO} G_k}{\sum_{k=1}^{k=3} G_k}$$

Example 2.15

Four electrical generators having e.m.f.s of 100, 110, 90, and 105 V with internal resistances of 5, 10, 2, and 12 Ω, respectively, are connected in parallel. Calculate the output voltage from the combination and also their combined output resistance.

Solution

Let

$$E_1 = 100 \text{ V}, G_1 = 1/5 = 0.2 \text{ S}$$
$$E_2 = 110 \text{ V}, G_2 = 1/10 = 0.1 \text{ S}$$
$$E_3 = 90 \text{ V}, G_3 = 1/2 = 0.5 \text{ S}$$
$$E_4 = 105 \text{ V}, G_3 = 1/12 = 0.083 \text{ S}$$

From Millman's theorem, the output voltage is

$$V_{O'O} = \frac{(100 \times 0.2) + (110 \times 0.1) + (90 \times 0.5) + (105 \times 0.083)}{0.2 + 0.1 + 0.5 + 0.083} = 95.94 \text{ V}$$

and the output resistance is

$$R_{out} = 1/\Sigma G = 1/0.883 = 1.13 \ \Omega$$

Note The above calculation provides the parameters for the Thévenin's circuit equivalent generator of the combination.

2.13 The General Star–Mesh Transformation (Rosen's Theorem)

It is possible to transform a star network of n components which are connected from a star point O' to n separate terminals (figure 2.29a) into a corresponding mesh network (figure 2.29b). In the mesh network, n conductances $G_{12}, G_{13}, G_{14}, \ldots, G_{1n}$, terminate on terminal (or node) 1. The equivalent conductance connected between the pair of terminals u and v in the network is

$$G_{uv} = \frac{G_u G_v}{\sum\limits_{k=1}^{k=n} G_k} \tag{2.32}$$

Equation 2.32 is a precise mathematical expression which describes the transformation, in which G_u is the conductance connected between terminal u and the star point, G_v is connected between terminal v and the star point, and ΣG_k is the sum of all the values of the conductances in the star system.

The transformation can be verified by applying Millman's theorem to the network, but in this case we connect terminal 1 to the reference node (terminal O in Millman's theorem) so that $V_1 = 0$, giving

$$V_{O'1} = \frac{V_{21} G_2 + V_{31} G_3 + \ldots}{G_1 + G_2 + G_3 + \ldots}$$

The current I_1 flowing in G_1 due to $V_{O'1}$ is

$$I_1 = V_{O'1} G_1 = \frac{V_{21} G_1 G_2 + V_{31} G_1 G_3 + \ldots}{G_1 + G_2 + G_3 + \ldots}$$

$$= V_{21} \frac{G_1 G_2}{G_1 + G_2 + G_3 + \ldots} + V_{31} \frac{G_1 G_3}{G_1 + G_2 + G_3 + \ldots} + \ldots$$

That is, the current flowing into terminal 1 is the sum of a number of currents, and the current flowing from terminal 2 to terminal 1 is due to a conductance of $G_1 G_2/\Sigma G$ connected between terminals 1 and 2. The component of current flowing to terminal 1 from terminal 3 is due to an equivalent conductance of $G_1 G_3/\Sigma G$ connected between terminals 1 and 3, etc. That is, the equivalent conductance linking the pair of terminals u and v is given by equation 2.32.

Only in the special case of three terminals is it possible to provide a unique transformation from a mesh circuit to a star circuit.

Example 2.16

Convert the star circuit in figure 2.30a into the mesh circuit shown in figure 2.30b.

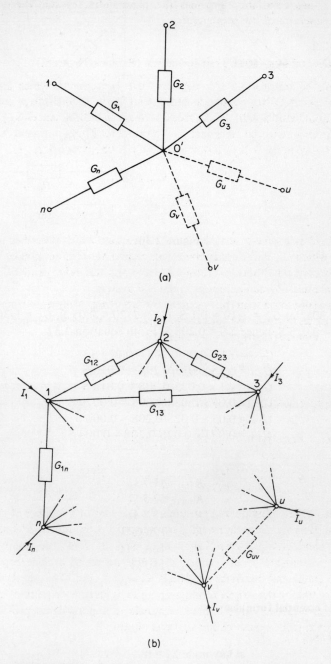

(a)

(b)

Figure 2.29 The general star–mesh transformation

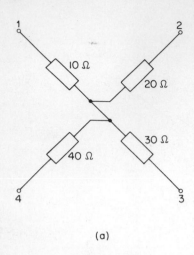

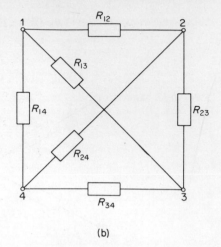

(a)   (b)

Figure 2.30

Solution

In figure 2.30a, let $G_1 = 1/10 = 0.1$ S, $G_2 = 1/20 = 0.05$ S, $G_3 = 1/30 = 0.033$ S, $G_4 = 1/40 = 0.025$ S, hence $\Sigma G = 0.208$ S. From equation 2.32

$$G_{12} = 0.1 \times 0.05/0.208 = 0.024 \text{ S}$$
$$G_{13} = 0.1 \times 0.033/0.208 = 0.016 \text{ S}$$
$$G_{14} = 0.1 \times 0.025/0.208 = 0.012 \text{ S}$$
$$G_{23} = 0.05 \times 0.033/0.208 = 0.0079 \text{ S}$$
$$G_{24} = 0.05 \times 0.025/0.208 = 0.006 \text{ S}$$
$$G_{34} = 0.033 \times 0.025/0.208 = 0.004 \text{ S}$$

Therefore

$$R_{12} = 41.7 \ \Omega$$
$$R_{13} = 62.5 \ \Omega$$
$$R_{14} = 83.3 \ \Omega$$
$$R_{23} = 126.6 \ \Omega$$
$$R_{24} = 166.7 \ \Omega$$
$$R_{34} = 250 \ \Omega$$

Summary of essential formulae

Kirchhoff's laws

$$\text{at any node } \Sigma I = 0$$
$$\text{in any mesh } \Sigma IR + \Sigma V = 0$$

Delta–star transformation (see figure 2.17)

$$R_{10} = R_{12}R_{13}/(R_{12} + R_{23} + R_{13})$$
$$R_{20} = R_{12}R_{23}/(R_{12} + R_{23} + R_{13})$$
$$R_{30} = R_{13}R_{23}/(R_{12} + R_{23} + R_{13})$$

Star–delta transformation (see figure 2.20)

$$R_{12} = R_{10} + R_{20} + R_{10}R_{20}/R_{30}$$
$$R_{23} = R_{20} + R_{30} + R_{20}R_{30}/R_{10}$$
$$R_{13} = R_{10} + R_{30} + R_{10}R_{30}/R_{20}$$

Millman's theorem

$$V_{O'O} = \frac{E_1 G_1 + E_2 G_2 + E_3 G_3}{G_1 + G_2 + G_3}$$

General star–mesh transformation

$$G_{uv} = G_u G_v/(G_1 + G_2 + \ldots + G_v)$$

PROBLEMS

2.1 Using Kirchhoff's laws, determine the current in each battery in figure 2.31. Calculate also the value of V_{XY} given that the e.m.f. and internal resistance of batteries A, B and C are, respectively, 108 V and 3 Ω, 120 V and 2 Ω, 30 V and 0 Ω. [$I_A = -0.183$ A, $I_B = 5.725$ A, $I_C = 5.542$ A; $V_{XY} = 108.6$ V]

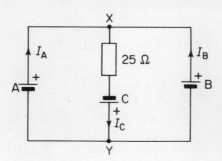

Figure 2.31

2.2 Solve problem 2.1 by using Thevenin's theorem and by using Norton's theorem.

2.3 Solve problem 2.1 using the superposition theorem.

2.4 Two batteries W and X are connected in parallel with one another, with like polarity terminals being connected together. The e.m.f. and internal resistance of W are 140 V and 10 Ω, respectively, and the corresponding values for X are 120 V and 4 Ω. A resistance of 25 Ω is connected between the battery terminals; calculate (a) the value and direction of the current in each battery, (b) the terminal voltage of the combination.

[(a) I_W = 2.718 A (discharging), I_X = 1.795 A (charging); (b) 112.82 V]

2.5 Six 2-Ω resistors are connected in a network between the following points: A-X, X-B, B-Y, Y-A, A-B, X-Y. A 10-V battery is connected in arm A-X with its positive terminal connected to X. Calculate the value of the current in branch X-Y, and the power consumed by the 2-Ω resistor in that branch.

[1.25 A; 3.125 W]

2.6 State the maximum power transfer theorem.

What should be the value of resistor R in figure 2.32 in order that it consumes the maximum power? Calculate the power consumed by R if E = 10 V.

[9 Ω; 1 W]

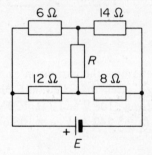

Figure 2.32

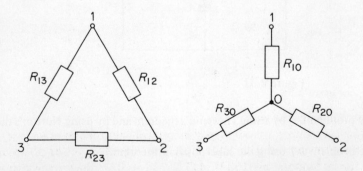

Figure 2.33

2.7 (a) If, in figure 2.33, $R_{12} = 12\ \Omega$, $R_{23} = 23\ \Omega$ and $R_{13} = 13\ \Omega$, calculate the value of the equivalent star-connected circuit R_{10}, R_{20} and R_{30}. (b) If $R_{10} = 10\ \Omega$, $R_{20} = 20\ \Omega$ and $R_{30} = 30\ \Omega$, calculate the values of the equivalent mesh-connected resistors R_{12}, R_{23} and R_{13}.

[(a) $R_{10} = 3.25\ \Omega$, $R_{20} = 5.75\ \Omega$, $R_{30} = 6.23\ \Omega$; (b) $R_{12} = 36.7\ \Omega$, $R_{13} = 55\ \Omega$, $R_{23} = 110\ \Omega$]

2.8 Transform the star-connected group of resistors between A, B and C in figure 2.34 into a mesh-connected network. Hence calculate the p.d. across the 60-Ω resistor.

[6 V]

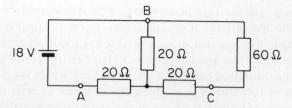

Figure 2.34

2.9 Write down the mesh equations for the circuit in figure 2.23 if the polarity of E_A is reversed. Solve the equations for I_1, I_2 and I_3 if $E_A = 100\ \text{V}$, $E_B = 80\ \text{V}$, and each resistor has a resistance of 100 Ω.

[$I_1 = -0.393\ \text{A}; I_2 = -0.18\ \text{A}; I_3 = -0.327\ \text{A}$]

2.10 For the circuit in figure 2.35, write down the node equations.

[$1 = 0.2V_A - 0.1V_B$; $10 = -0.1V_A + 0.7V_B$]

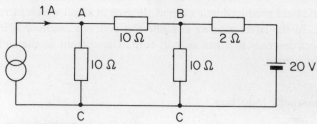

Figure 2.35

2.11 State Millman's theorem.

Three d.c. generators are connected in parallel with one another, the output terminals being connected to a 50-Ω load. The respective values of open-circuit e.m.f. of the generators are 100 V, 115 V and 120 V; their respective internal resistances are 10, 12.5 and 15 Ω. Calculate the current flowing in the 50-Ω load.

[2.115 A]

3 Electromagnetism

3.1 Magnetic fields

Since we do not fully understand what a magnetic field is, we simply say that it is a *condition of space*. The presence of a magnetic field is detected by its effects, such as those on an electric current, or on iron filings, or on a permanent magnet. The *direction* in which the magnetic field acts at a particular point is given by the direction in which the force would act on an isolated north-seeking pole (a N-pole) at that point. The theoretical concept of an isolated north pole is one which is used simply to define certain aspects of magnetic fields. When we refer to a N-pole, we imply that the pole experiences a force in the earth's magnetic field towards the earth's north pole. Similarly, a S-pole is a south-seeking pole.

If our isolated N-pole were free to move in space, it would move in the direction of the force acting on it, that is, in the direction of the magnetic field, and would trace out a *line of magnetic flux*. This line of flux is, in fact, stationary but in some instances it is convenient to assume that the flux 'flows' around the magnetic circuit.

A *magnetic circuit* is simply an interconnected set of branches in which a magnetic flux is established, and has an almost exact analogy with the electrical circuit (see also section 3.16). The magnetic flux that is established in a magnetic material can occur as a result of either (i) permanent magnetism in the material or (ii) a magnetic field produced by a current flowing in a coil which surrounds part of the material. In the latter case we describe the coil as a *solenoid* if it is air-cored, and as an *electromagnet* if it has an iron core. The current in the coil is known as the *excitation current*.

3.2 Electromagnetic Induction

When a current flows in a solenoid a magnetic flux is established in the magnetic circuit; the larger the value of excitation current, the larger the value of magnetic flux. The converse of this is also true, that is, **if the flux linking with a solenoid is altered then an e.m.f. is induced in the coil, and would cause a current to flow in the coil if the electrical circuit is complete.** This is the basis of electromagnetic induction.

The induced e.m.f. can be due to a number of causes. If, for example, the exciting current in the coil is increased, then the magnetic flux linking with the coil increases. This causes an e.m.f. to be induced in the coil *due to the change of current,* and is known as a *self-induced e.m.f.* If the induced e.m.f. is due to the change of flux in another coil which is magnetically linked with or magnetically coupled to the first coil, it is known as a *mutually induced e.m.f.*; the operation of transformers is based on the fact that e.m.f.s are induced in mutually coupled coils. The e.m.f. can also be induced as a result of relative movement between the coil and the magnetic field, and is known as *induction by motion*, and is the basis of the electrical generator.

These effects are summarised by the laws of electromagnetism, which are given below.

Faraday's law

The magnitude of the induced e.m.f. is proportional to the rate of change of the magnetic flux linking the circuit.

Lenz's law

The induced e.m.f. acts to circulate a current in a direction that opposes the change in flux which caused the induced e.m.f.

3.3 Magnetic Flux and Flux Density

Resulting from the laws stated in section 3.2, the e.m.f. e induced in a coil of N turns when the flux linking with it is changing is

$$e \propto N \frac{d\Phi}{dt} \tag{3.1}$$

where $d\Phi$ is the change in magnetic flux linking the circuit. The unit of flux is the *weber*, Wb. The dimensions of the quantities in equation 3.1 used in the SI system are chosen so that the coefficient of proportionality in equation 3.1 is unity, and

$$e = N \frac{d\Phi}{dt} \tag{3.2}$$

In some instances, the electrical transformer (see chapter 9) being an example, the polarity of the e.m.f. induced in an electrically isolated winding depends on the terminal selected as the reference node in the circuit. In such cases the induced e.m.f. may be of the opposite polarity to that applied to the primary winding which produces the flux. In this case equation 3.2 is re-written as

$$e = -N\frac{\mathrm{d}\varPhi}{\mathrm{d}t} \qquad (3.3)$$

Equations 3.2 and 3.3 are sometimes combined in the following form

$$e = \pm\frac{\mathrm{d}(N\varPhi)}{\mathrm{d}t} \qquad (3.4)$$

where the term $N\varPhi$ is known as the *magnetic flux linkage* (Wb turns or Wb).

The equations given above give the *instantaneous value* of the induced e.m.f. The *average value* of induced e.m.f. can be calculated as follows. Suppose that flux $\varPhi_1$ links with a circuit at time t_1, and its value changes to $\varPhi_2$ in a time interval $t_2 - t_1$ (figure 3.1). The change in flux $\mathrm{d}\varPhi = (\varPhi_2 - \varPhi_1)$ occurs in time $\mathrm{d}t = (t_2 - t_1)$, and the average e.m.f. E induced in the coil is

$$E = N\frac{(\varPhi_2 - \varPhi_1)}{(t_2 - t_1)} \qquad (3.5)$$

The mechanism of inducing an e.m.f. is illustrated in figure 3.1. If the flux changes along line A in figure 3.1a, the rate of change of flux is uniform. The net result is

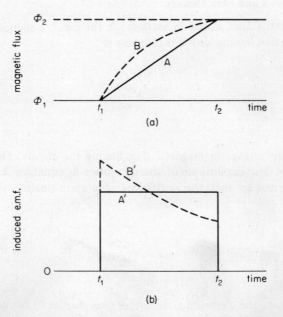

(a)

(b)

Figure 3.1   The relationship between the rate of change of magnetic flux and the induced e.m.f.

that the induced e.m.f. has a constant value during the time interval between t_1 and t_2. On the other hand if the flux change is non-linear as in curve B in figure 3.1a, then the curve of the instantaneous value of induced e.m.f. in the circuit is also non-linear, shown in curve B' in figure 3.1b. Nevertheless, since the flux change is the same in the cases of curves A and B, then the *average* value of the induced e.m.f. is the same in both cases.

The *magnetic flux density*, B, is the amount of flux passing through unit area and has the unit of the *tesla*, T, where

$$B = \frac{\text{magnetic flux}}{\text{area perpendicular to the flux}} = \frac{\Phi}{A} \quad \text{tesla}$$

where Φ is the magnetic flux in webers and A is the area in m^2 perpendicular to the flux path through which the flux passes. Thus one tesla is equivalent to a density of one Wb/m^2.

Example 3.1

The flux linking an air-cored coil of 500 turns changes from 30 μWb to 60 μWb in 2 ms. Calculate the value of the e.m.f. induced in the coil.

Solution

$$d\Phi = 60 - 30 = 30 \ \mu\text{Wb}$$

$$dt = 2 \text{ ms}$$

hence

$$e = N\frac{d\Phi}{dt} = \frac{500 \times 30 \times 10^{-6}}{2 \times 10^{-3}} = 7.5 \text{ V}$$

Example 3.2

The flux density in a magnetic circuit of area 300 cm^2 is 0.01 T. Calculate the value of the magnetic flux in the circuit.

Solution

From the relationship $B = \phi/A$ then

$$\Phi = A B = 300 \times (10^{-2})^2 \times 0.01 = 300 \times 10^{-6} \text{ Wb}$$

$$= 300 \ \mu\text{Wb or } 0.3 \text{ mWb}$$

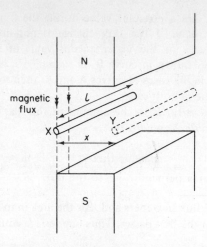

Figure 3.2   The e.m.f. induced in a single conductor when it cuts magnetic flux

3.4 Induced E.M.F. due to Motion

Suppose that the conductor shown in figure 3.2, having an active length l in the magnetic field, is moved through distance x in t seconds. If the flux density in the air gap is B tesla, then the flux cut by the conductor when it moves from X to Y is

$$\text{change in flux} = B \times \text{area} = Blx \quad \text{Wb}$$

and if the linear velocity of the conductor is v m/s, then

$$v = x/t \quad \text{m/s}$$

From equation 3.5, the average value of e.m.f. induced in the conductor (which is equivalent to part of a coil of a single turn) is

$$E = \frac{\text{change in flux}}{\text{time } t} = \frac{Blx}{x/v} = Blv \quad \text{volts} \tag{3.6}$$

Should the direction of motion of the conductor be at angle θ to the direction of the magnetic field as shown in figure 3.3, then the rate at which the flux lines are cut is reduced when compared with the case where the conductor movement is perpendicular to the field. The equation for the average induced e.m.f. then becomes

$$E = Blv \sin \theta \tag{3.7}$$

Readers will note from equation 3.7 that when the conductor moves *in the same direction* as the flux lines ($\theta = 0$), no flux is cut and the induced e.m.f. is zero.

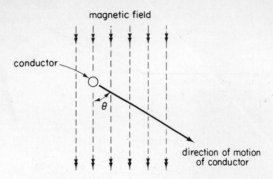

Figure 3.3   The effect of moving a conductor at an angle to the field

Example 3.3

A conductor of active length 0.3 m moves perpendicular to a magnetic field at a velocity of 50 m/s. Calculate the average value of the e.m.f. induced in the conductor if the magnetic flux density is 0.5 T.

Solution

$$E = Blv = 0.5 \times 0.3 \times 50 = 7.5 \text{ V}$$

Example 3.4

The average e.m.f. induced in a conductor of length 0.6 m is 0.5 V. The conductor moves at an angle to a magnetic field of flux density 0.3 T at a velocity of 10 m/s. Calculate the value of the angle between the direction of movement of the conductor and the direction of the magnetic field.

Solution

$$E = Blv \sin \theta$$

hence

$$\sin \theta = E/Blv = 0.5/(0.3 \times 0.6 \times 10) = 0.2778$$

or

$$\theta = 16.13°$$

3.5 Fleming's Right-hand Rule

The direction in which the induced e.m.f. acts in a conductor can be deduced by the use of Fleming's right-hand rule as follows.

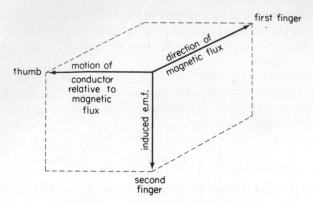

Figure 3.4   Fleming's right-hand rule

If the thumb, first finger and second finger of the right hand are held so that they point in directions that are mutually perpendicular (see figure 3.4) **then if the *F*irst finger points in the direction of the magnetic *F*lux, and the thu*M*b points in the direction of the *M*otion of the conductor, then the s*E*cond finger points in the direction of the induced *E*.m.f.**

That is

> *F*irst finger − direction of magnetic *F*lux
> s*E*cond finger − direction of *E*.m.f.
> thu*M*b − *M*otion of conductor *relative* to the flux

To illustrate the application of this rule we will consider the electro-mechanical systems in figure 3.5. In figure 3.5a the conductor moves to the left relative to the magnetic field, and the direction of the magnetic flux is from the N-pole to the S-pole. Applying Fleming's right-hand rule we see that the direction of the induced

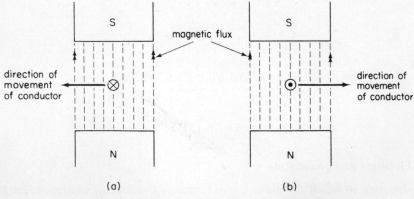

Figure 3.5   Application of Fleming's right-hand rule

e.m.f. would be such as to cause a current to flow *into* the paper. Since current flow is represented by means of an arrow, we indicate that current is flowing away from the reader (that is, into the paper) by showing the 'tail' or 'crossed feathers' of the arrow on the conductor.

Applying Fleming's right-hand rule to figure 3.5b indicates that the induced e.m.f. acts in a direction to cause current to flow *out of* the paper. This fact is indicated by the dot in the centre of the conductor, which represents the 'point' of the arrow approaching the reader.

3.6 Direction of the Magnetic Field around a Conductor

It was stated in section 3.1 that the direction of the magnetic field at any point is given by the direction in which an isolated N-pole would move if placed at that point. Since an isolated pole is a theoretical concept, experiments are often carried out with compass needles and, in the case of a straight cylindrical conductor, the magnetic flux lines are found to follow a circular path around the conductor in the manner shown in figure 3.6a. Moreover, the direction of the magnetic field is found to be related to the direction of flow of the current. Reversing the direction of flow of current reverses the direction of the magnetic field.

A simple and convenient method of relating the directions of magnetic field to that of the current is given by the *screw rule*. If we imagine the wood screw in figure 3.6b to point in the direction of flow of current, then in order to propel the

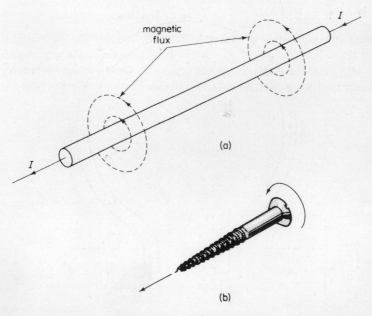

Figure 3.6 The screw rule

screw forward in the direction of the current, we turn the screw head in a clockwise direction, that is, in the same direction as the magnetic field.

Alternatively, we may use a method known as the *right-hand screw rule* which states that if the conductor is grasped with the right hand with the thumb outstretched parallel to the conductor and pointing in the direction of the flow of current, then the fingers point in the direction of the magnetic field around the conductor.

3.7 Flux Distribution in a Coil

In the case of a single-turn coil (figure 3.7) the flux distribution along section AB is as shown in the plan view in the upper part of the diagram. Since the current flows

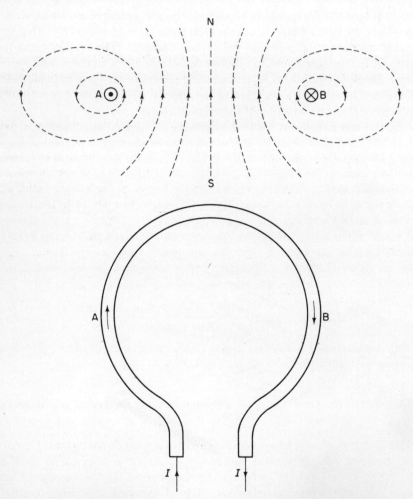

Figure 3.7   Magnetic field produced by a current flowing in a single turn of wire

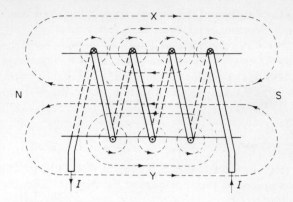

Figure 3.8   Magnetic field produced by the current in a solenoid

upwards at point A, then the flux path at that point is anticlockwise. At point B the direction of the current is downwards, so that the flux path follows a clockwise path. The net result, shown in the figure, is that a N-pole is formed on one side of the coil, and a S-pole is formed on the other side.

The flux pattern associated with a *solenoid* or a multi-turn coil can be deduced from the foregoing. A section through a solenoid is shown in figure 3.8. The resultant flux pattern is deduced by the application of the rules given above, and we see that the bulk of the magnetic flux leaves the left-hand end of the solenoid and enters the right-hand end. In this way a N-pole is formed at the left-hand end, and a S-pole at the right-hand end. Not all the flux follows the main path, illustrated in the cases of paths X and Y.

In many applications the *useful* magnetic flux is that which either enters or leaves the ends of the solenoid. The flux which fails to follow the 'useful' path is known as *magnetic leakage* or *fringing*, and is accounted for in calculations by means of a factor known as the *leakage coefficient* as follows

$$\text{leakage coefficient} = \frac{\text{total magnetic flux produced}}{\text{useful magnetic flux}}$$

In a well-designed magnetic system, for example, an electrical machine, the value of this coefficient may lie in the range 1.15 to 1.25.

3.8 Direction of the Force Acting on a Current-carrying Conductor in a Magnetic Field

The flux distribution around an isolated current-carrying conductor takes the form shown in figure 3.9a. When the conductor is placed in a magnetic field (see figure 3.9b), the flux lines produced by the current in the conductor *assist* the main magnetic field on the right-hand side of the conductor, and oppose it to the left of the conductor. The net result is a distortion of the magnetic field, with

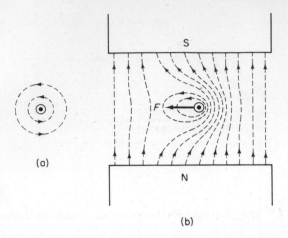

(a)

(b)

Figure 3.9   Force acting on a current-carrying conductor in a magnetic field

the flux being increased on the right-hand side of the conductor and reduced on the left-hand side.

In this situation the conductor experiences a mechanical force F acting to displace it from the region where the field is strongest to where it is weakest. That is, from right to left in the figure.

This is the basis of the *electrical motor*, and is also the principle upon which the electromagnetic deflection of cathode rays in television tubes is based. In the latter case the current is simply a beam of electrons in motion, the amount by which 'current elements' are deflected depends on the strength of the magnetic field.

From figure 3.9 readers will note that the directions of the electric current, the magnetic flux and the mechanical force are mutually perpendicular to one another. In any particular case their relative directions can be deduced by use of *Fleming's left-hand rule*† as follows

If the thumb, first finger and second finger of the left hand are held so that they point in directions which are mutually perpendicular to one another (see figure 3.10), then if the *F*irst finger points in the direction of the magnetic *F*lux and the se*C*ond finger points in the direction of the flow of *C*urrent, then the thu*M*b points in the direction of the force acting on the conductor, that is, it gives the direction of the *M*otion of the conductor.

That is

*F*irst finger — direction of magnetic *F*lux
se*C*ond finger — direction of *C*urrent
thu*M*b — direction of *M*otion of the conductor

†To recall that Fleming's left-hand rule applies to motor action, simply remember that **motors** drive on the **left-hand** side of the road in Britain.

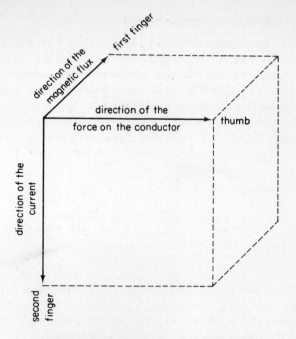

Figure 3.10 Fleming's left-hand rule

3.9 Magnitude of the Force on a Current-carrying Conductor in a Magnetic Field

It has been determined experimentally that the force F experienced by the conductor in figure 3.9 is given by the expression

$$F = BIl \quad \text{newtons} \tag{3.8}$$

where B is the flux density in tesla, I is the current in the conductor in amperes, and l is the active length of the conductor, in metres, in the magnetic field.

Example 3.5

A conductor carries a current of 50 A in a flux density of 0.25 T, the length of the conductor being perpendicular to the magnetic field. If the active length of the conductor is 0.6 m, calculate the value of the force acting on the conductor.

Solution

$$F = BIl = 0.25 \times 50 \times 0.6 = 7.5 \text{ N}$$

3.10 Force between Parallel Current-carrying Conductors

In the case of conductors that carry currents in opposite directions (see figure 3.11a), the force F_A acting on conductor A is due to the interaction between the current in A and the magnetic flux produced by the current in B. Similarly, force F_B acting on conductor B is due to the interaction between the current in B and the flux produced by the current in conductor A. Applying Fleming's left-hand rule to each of the conductors in figure 3.11a, we deduce that the directions of the forces are as shown and that there is a force of repulsion between the two conductors.

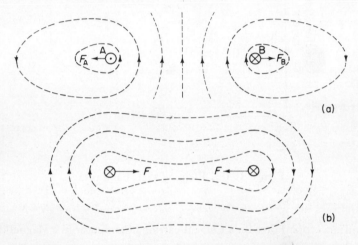

Figure 3.11   Force acting between current-carrying conductors

Where the conductors carry currents in the same direction, figure 3.11b, the force between the conductors is one of attraction.

3.11 Magnetomotive Force (F) and Magnetic Field Intensity (H)

The magnetomotive force (m.m.f.) is the force that causes a magnetic flux to be established. Comparing magnetic and electrical circuits, magnetic flux and electrical current are analogous quantities, as are the m.m.f. and e.m.f. in magnetic and electrical circuits respectively.

The m.m.f., symbol F, is proportional to the current flowing in the circuit or coil, and is also proportional to the number of turns, N, on the coil; hence

$$F = IN \quad \text{ampere turns (At) or amperes} \tag{3.9}$$

Since N is a number or numeric and has no dimensions, then F may be given the dimensions of amperes.

The *magnetic field intensity* (known also as the *magnetising force* or the *magnetic field strength*), symbol H, is the m.m.f. per unit length, as follows

$$H = F/l = IN/l \quad \text{At/m or A/m} \tag{3.10}$$

where l is the length of the magnetic circuit in metres.

Example 3.6

A 200-turn coil carries a current of 5 A. If the length of the coil is 0.1 m, calculate (i) the m.m.f. produced by the coil and (ii) the magnetic field intensity inside the coil.

Solution

(i) $F = IN = 5 \times 200 = 1000$ At or A
(ii) $H = F/l = 1000/0.1 = 10\ 000$ At/m or A/m

3.12 Permeability

In a given magnetic medium, the flux density B is related to the magnetising force H which produces it by the relationship

$$B = \mu H \tag{3.11}$$

where μ is the *absolute permeability* of the material, and has the dimensions of henrys/metre (H/m). The *permeability of free space* or *magnetic space constant* is given a special symbol, μ_0, and has the value

$$\mu_0 = 4\pi \times 10^{-7} \quad \text{H/m} \tag{3.12}$$

If a ferromagnetic core is inserted inside the former of an air-cored coil the flux density is intensified, the factor by which it is increased is given by the *relative permeability* μ_r of the material, where

$$\mu_r = \frac{\text{flux density with the ferromagnetic core}}{\text{flux density without the ferromagnetic core}}$$

$$= \frac{\mu H}{\mu_0 H} = \frac{\mu}{\mu_0} \tag{3.13}$$

Inserting equation 3.13 into equation 3.11 yields

$$B = \mu_r \mu_0 H \tag{3.14}$$

3.13 Classification of Magnetic Materials

Materials can broadly be classified as being either magnetic or non-magnetic, but between the two extremes lie a number of groups of magnetic materials. The following is a summary of the more important groups.

Ferromagnetic materials

The permeability of these elements is considerably greater than that of a vacuum. These materials include iron, steel, nickel and cobalt as well as a number of alloys, for example, nickel—iron and cobalt—iron. The relative permeabilities of these elements vary with magnetising force, and have a peak in their relative-permeability—magnetising-force characteristics.

Paramagnetic materials

These materials have a relative permeability whose value is slightly greater than unity, and become weakly magnetised in the direction of the magnetising field. Included in this group are aluminium, chromium, manganese and platinum.

Diamagnetic materials

The relative permeability of these materials is less than unity, and they become weakly magnetised but in the opposite direction to that of the magnetising field. Diamagnetic materials include antimony, bismuth, copper, gold, silver and zinc. To illustrate the value of relative permeability involved, in the case of bismuth it is 0.999 82.

Ferrites

These are semiconductor materials that have similar ferromagnetic properties to iron but, like many semiconductors, have a high resistivity. The latter property leads to a very low power loss at high frequencies.

3.14 Magnetisation Curves of Ferromagnetic Materials

The application of a steadily increasing magnetising force to a piece of ferromagnetic material which was originally demagnetised causes the flux density to increase in the manner shown in figure 3.12.

Between the origin O and point K the flux density increases rapidly, but beyond point K the slope of the curve decreases. Point K indicates the onset of magnetic saturation. Increasing the magnetising force further results in an increase in flux density but at a progressively lower rate until, at S, the increase in flux with H is only equal to the same increase that would result if the coil were air-cored. When this occurs, the core material is said to be *magnetically saturated*.

The shape of the curve in figure 3.12 can be explained in terms of the physics of ferromagnetic materials as follows. The magnetic properties of ferromagnetic materials depend on the magnetic field associated with electrons in motion. The spinning motion of electrons produces a tiny permanent-magnet effect, and two

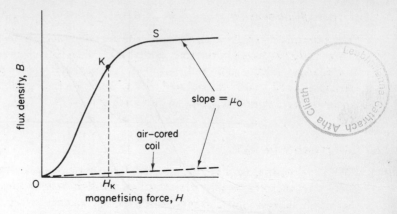

Figure 3.12 Magnetisation curve of a ferromagnetic material

electrons spinning in opposite directions form a non-magnetic pair. In the molecular structure of ferromagnetic materials, atoms with like spins are grouped together to form what are called *domains* or *dipole magnets*. In demagnetised materials the domains point in different directions and the net magnetic flux density is zero. Under the influence of an externally produced magnetic field, the domains line up so that the magnetic flux density increases very rapidly. For values of magnetising force in figure 3.12 greater than H_K, the number of domains remaining to be lined up with the field is reduced until, at point S all the domains are in line with the magnetising field.

When the magnetising force is reduced to zero, some domains return to their original directions and some do not. As a result, the material retains some of its magnetism and is a measure of the *retentivity* or *residual magnetism* of the material. This is indicated in figure 3.13 by the *remanent flux*, B_r. In materials used for permanent magnets, a high retentivity is desirable. The residual flux is reduced to zero by applying a reverse magnetising force, the *coercive force H_c* in figure 3.13.

If the reverse magnetising force is increased further, the material becomes saturated once more at point X in figure 3.13, but with the opposite magnetic polarity. The complete loop in figure 3.13 is known as a *hysteresis loop* or *B–H loop*.

It can be shown that the area of the *B–H* loop has dimensions of energy per unit volume, that is, J/m^3, so that every time the magnetising force suffers a complete cycle, that is, two reversals, power is consumed by the material. This power loss is known as the *hysteresis loss*, P_h, and is related to the frequency of reversals, f, and also to the maximum flux density B_{max} as follows

$$\text{hysteresis loss} \propto f(B_{max})^n$$

where n is a number whose value lies between 1.6 and 2.2, and is typically 1.7.

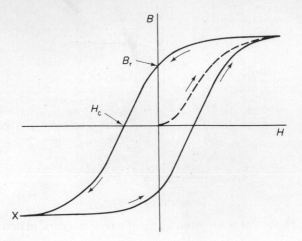

Figure 3.13   *B–H* curve for ferromagnetic material

3.15 Magnetic Circuits

The relationship between the m.m.f., F, applied to a magnetic circuit and the flux Φ in the circuit is given by the relationship

$$F = \Phi S \quad \text{At} \tag{3.15}$$

where S is the *reluctance* of the circuit, and has dimensions of At/Wb or A/Wb. Hence

$$S = F/\Phi \quad \text{At/Wb}$$

Now

$$F = Hl$$

and

$$\phi = Ba = \mu_r \mu_0 \, HA$$

where l is the length of the magnetic circuit and A is its area. Therefore

$$S = \frac{Hl}{\mu_r \mu_0 \, HA} = \frac{l}{\mu_r \mu_0 \, A} \quad \text{A/Wb} \tag{3.16}$$

In some cases it is convenient to use a quantity known as the *permeance*, symbol Λ, of the magnetic circuit, where

$$\Lambda = 1/S \quad \text{Wb/A or Wb/At} \tag{3.17}$$

Equation 3.15 is analogous to the expression $E = IR$ for the electrical circuit. Also equation 3.16 is generally similar to $R = \rho l/A$ for the electrical circuit. A comparison between electrical and magnetic circuit quantities is given in table 3.1.

Table 3.1

Magnetic circuit			Electrical circuit		
Quantity		Unit	Quantity		Unit
m.m.f.	F	At	e.m.f.	E	V
magnetising force	H	At/m	electric force	E	V/m
magnetic flux	Φ	Wb	electric current	I	A
reluctance	S	A/Wb	resistance	R	Ω
permeance	Λ	Wb/A	conductance	G	S

$$F = \Phi S$$

$$S = \frac{l}{\mu_r \mu_0 A}$$

$$E = IR$$

$$R = \rho \frac{l}{A}$$

Example 3.7

A mild-steel ring has a mean circumference of 0.25 m and cross-sectional area of 0.001 m². If the flux in the core is 1.5 mWb, and the relative permeability of the ring at the operating flux density is 625, calculate the current required in a coil of 1000 turns wound uniformly on the core to produce this flux.

Solution

$$S = l/(\mu_r \mu_0 A) = 0.25/(625 \times 4\pi \times 10^{-7} \times 0.001)$$
$$= 0.3183 \times 10^6 \text{ A/Wb}$$

m.m.f. $F = \Phi S = 1.5 \times 10^{-3} \times 0.3183 \times 10^6 = 478$ At

hence

$$I = F/N = 478/1000 = 0.478 \text{ A}$$

3.16 Magnetic Circuits in Series and in Parallel

The analogy between electrical and magnetic circuits is sufficiently close to allow us to use techniques in the solution of magnetic circuits that are similar to those used in electrical circuits. Consider the magnetic circuit in figure 3.14a, which consists of an iron path of length l_1 and an air gap of length l_2 in series with one another. The 'equivalent' circuit is shown in figure 3.14b, in which

$$F = NI$$
$$S_1 = l_1/\mu_r \mu_0 A$$

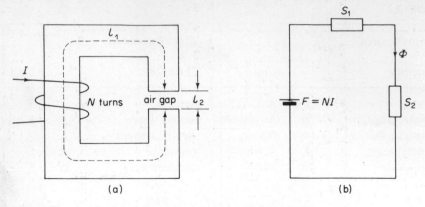

Figure 3.14   A series magnetic circuit

$$S_2 = l_2/\mu_0 A$$

$$\Phi = F/(S_1 + S_2)$$

Figure 3.15a illustrates a series–parallel magnetic circuit. The values of the circuit elements in the equivalent circuit, figure 3.15b, are calculated as follows

$$F = NI$$

$$S_1 = l_1/\mu_{r1}\,\mu_0\,A_1$$
$$S_2 = l_2/\mu_{r2}\,\mu_0\,A_2$$
$$S_3 = l_3/\mu_{r3}\,\mu_0\,A_3$$
$$S_4 = l_4/\mu_0\,A_3$$
$$S_5 = l_5/\mu_{r5}\,\mu_0\,A_5$$

where a_1, a_2, and a_3 are the areas of the left-hand, centre and right-hand limbs, respectively, of the circuit and μ_{r1}, μ_{r2}, etc. are the relative permeabilities of the iron sections at their working flux densities. Applying the rules of series–parallel electrical circuits to calculate the equivalent reluctance, S, of the circuit in figure 3.15b yields

$$S = S_1 + \frac{S_2(S_3 + S_4 + S_5)}{S_2 + S_3 + S_4 + S_5}$$

and

$$\Phi = F/S$$

also

$$\Phi_1 = \Phi \times (S_3 + S_4 + S_5)/(S_2 + S_3 + S_4 + S_5)$$

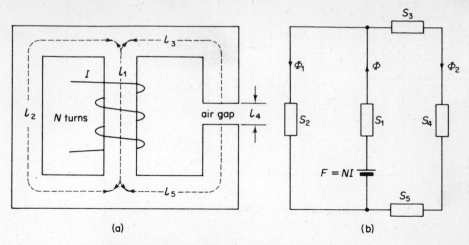

Figure 3.15   A series–parallel magnetic circuit

and

$$\Phi_2 = \Phi \times S_2 / (S_2 + S_3 + S_4 + S_5)$$

Example 3.8

A magnetic circuit of the type in figure 3.14a has an iron part with the B–H curve shown in figure 3.16. If the length of the iron part is 50 cm and the length of the air gap is 1 mm, calculate the value of current required in a coil of 300 turns wound uniformly around the iron core to produce a flux of 800 μWb in the core. The cross-sectional area of the magnetic circuit is 5 cm^2.

Solution

$$\text{Flux density} = B = \Phi/A = 800 \times 10^{-6}/5 \times 10^{-4} = 1.6 \text{ T}$$

From figure 3.16 we note that to produce this flux density in the iron circuit, a magnetising force of 3500 At/m is required. The number of ampere turns required to produce a flux density of 1.6 T in the iron circuit of length 50 cm is

$$F_1 = 3500 \times 0.5 = 1750 \text{ At}$$

The number of ampere turns F_2 required to produce the flux in the air gap is

$$F_2 = \Phi S_2 = \Phi l_2/\mu_0 A$$
$$= 800 \times 10^{-6} \times 1 \times 10^{-3}/(4\pi \times 10^{-7} \times 5 \times 10^{-4})$$
$$= 1274 \text{ At}$$

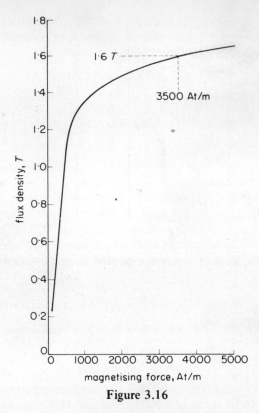

Figure 3.16

The total ampere-turn requirement for the complete circuit is

$$F = F_1 + F_2 = 1750 + 1274 = 3024 \text{ At}$$

Hence the current I that must flow in the coil is

$$I = F/N = 3024/300 = 10.08 \text{ A}$$

The effect of magnetic leakage on magnetic circuit calculations

Suppose in the above problem that the magnetic leakage coefficient is 1.05, that is, the magnetic flux in the iron section is 5 per cent higher than in the air gap. This implies that 5 per cent of the total flux bypasses the air gap through leakage paths. In this event the flux density in the iron is

$$\Phi/A = 1.05 \times 800 \times 10^{-6}/(5 \times 10^{-4}) \doteqdot 1.68 \text{ T}$$

From figure 3.16, the estimated magnetising force to produce this flux density is 6000 At/m. Hence

$$F_1 = 6000 \times 0.5 = 3000 \text{ At}$$

Since the flux in the air gap is unchanged at 800 μWb, the value of F_2 remains 1274 At. Hence the total ampere-turn requirement to produce an air gap flux of 800 μWb when the leakage coefficient is 1.05 is

$$F = F_1 + F_2 = 3000 + 1274 = 4274 \text{ At}$$

The current in the coil is therefore

$$I = F/N = 4274/300 = 14.25 \text{ A}$$

3.17 Self Inductance, L

It was shown in section 3.2 that an e.m.f. is induced in a coil if the flux produced by that coil changes, that is, the e.m.f. is induced when the current in the coil changes. The unit of self inductance, the *henry* (symbol H), is defined as follows.

A circuit has a self inductance of one henry if an e.m.f. of one volt is induced in the circuit when the current in the circuit changes at the rate of one ampere per second. Hence

self-induced e.m.f. $= e = L \times$ rate of change of current

$$= L\frac{di}{dt} \quad \text{volts} \tag{3.18}$$

It was also shown in section 3.3 that the induced e.m.f. is given by the equation

$$e = d(N\Phi)/dt$$

Therefore

$$L\frac{di}{dt} = \frac{d(N\Phi)}{dt}$$

or

$$L = \frac{d(N\Phi)}{di} = \frac{\text{change in magnetic flux linkages}}{\text{change in current}} \quad \text{H} \tag{3.19}$$

For a magnetic circuit with constant reluctance, the flux is proportional to the exciting current, hence if a flux Φ is produced by current I, then

$$L = N\Phi/I$$

Now

$$\Phi = BA = \mu HA = \mu\frac{NI}{l}A$$

or

$$L = \mu N^2 \, A/l = N^2/S \tag{3.20}$$

Hence the inductance of a coil is proportional to N^2, and doubling the number of turns on the coil quadruples its inductance.

Example 3.9

Calculate the average value of self-induced e.m.f. in a coil of 0.5 H when the current flowing through it is increased from 0.1 A to 2.1 A in 50 ms.

Solution

$$di = 2.1 - 0.1 = 2 \text{ A}$$

$$dt = 50 \text{ ms}$$

Hence the average value of the self-induced e.m.f. is

$$E = L \, di/dt = 0.5 \times 2/(50 \times 10^{-3}) = 20 \text{ V}$$

Example 3.10

A coil of 500 turns has an inductance of 15 mH. Determine (i) the flux produced in the core when a current of 2 A flows in the coil and (ii) the value of the self-induced e.m.f. in the coil when the current is changed from +2 A to −2 A in 10 ms.

Solution

(i) From equation 3.19

$$L = \frac{\text{change in flux linkages}}{\text{change in current}} = \frac{N(\Phi_2 - \Phi_1)}{I_2 - I_1}$$

where Φ_1 and I_1 are the initial values of magnetic flux and current, respectively, and Φ_2 and I_2 are the final values. If the initial values of flux and current are zero, then

$$L = N\Phi_2/I_2$$

Therefore

$$\Phi_2 = LI_2/N = 15 \times 10^{-3} \times 2/500 \text{ Wb} = 60 \, \mu\text{Wb}$$

(ii) Since the current changes from +2 to −2 A in a period of 10 ms, then

$$di = I_2 - I_1 = -2 - (+2) = -4 \text{ A}$$

Therefore

$$di/dt = -4/10 \times 10^{-3} = -400 \text{ A/s}$$

From equation 3.18

$$\text{induced e.m.f.} = L \, di/dt = 15 \times 10^{-3} \times (-400) = -6 \text{ V}$$

3.18 Mutual Inductance, M

Two coils are said to be mutually coupled when a change in the magnetic flux produced by one coil causes an e.m.f. to be induced in the other. For this to occur, the flux produced by either coil must link or *cut* the windings of the other coil. This is the basis of the operation of electrical transformers, and the coils may either be *closely coupled* when the majority of the flux links with both coils, or be *loosely coupled* when only a small proportion of the flux produced links with both coils. The degree of coupling between the coils is indicated by the value of the *coupling coefficient*, symbol k, which has a maximum value of unity for coils that are closely coupled, and has zero value when the coils have no magnetic coupling between them.

. Let us consider the mutually coupled coils in figure 3.17. When switch S is closed, the current in the *primary winding* (coil A) flows in the direction shown, and while the current is increasing from zero to its final value, the magnetic flux produced by the coil also increases. As shown in equation 3.2, an e.m.f. e_2 is

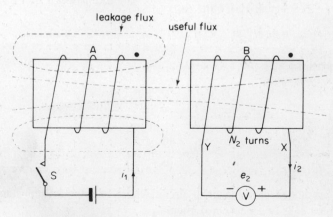

Figure 3.17   Mutual inductance between coils

induced in the *secondary winding* (coil B) *during the time that flux* $_1$ *is changing*.

The polarity of the mutually induced e.m.f. in the secondary winding in figure 3.17 can be deduced from a consideration of Lenz's law. This states that the induced e.m.f. acts in a direction which would cause a current to circulate so that the flux it produces would oppose the flux change producing the e.m.f. In the case considered, the flux enter the left-hand end of the coil B and is increasing in magnitude as the primary current increases. If the secondary induced e.m.f. were to cause a current to circulate, it would produce a magnetic flux opposing the flux entering coil B. Applying the rules described earlier, we see that the current through coil B from terminal Y to terminal X, that is, the current in the external circuit, flows from X to Y. Hence, terminal X is positive with respect to terminal Y.

3.19 Magnetic Coupling Coefficient, k

The flux Φ_1 in figure 3.17 produced by i_1 in coil A consists of two components, namely the useful flux Φ_{12} (pronounced phi one, two) linking coils A and B, and the leakage flux Φ_{11} (pronounced phi one, one), which links only with coil A. Thus

$$\Phi_1 = \Phi_{11} + \Phi_{12}$$

where Φ_{12} is the flux which induces e.m.f. e_2 in coil B. By Faraday's law of electromagnetic induction

$$e_2 = N_2 \; d\Phi_{12}/dt \qquad (3.21)$$

Also

$$e_2 = \frac{\text{mutual inductance } (M)}{\text{between the coils}} \times \text{rate of change of } i_1$$

$$= M \; di_1/dt \qquad (3.22)$$

Equating equations 3.21 and 3.22 gives

$$M \; di_1/dt = N_2 \; d\Phi_{12}/dt$$

hence

$$M = N_2 \; \frac{d\Phi_{12}}{di_1} \qquad (3.23)$$

If the coils are linked by a magnetic circuit whose reluctance is constant (as in the case of coupled coils on an air core), then

$$M = N_2 \; \frac{\Phi_{12}}{I_1} \qquad (3.24)$$

If the roles of coils A and B are reversed so that coil B is excited by the supply we find that

$$M = N_1 \; \frac{d\Phi_{21}}{di_2} \qquad (3.25)$$

where Φ_{21} is the part of the flux produced by coil B which links with coil A. Also, if a constant-reluctance core is used then

$$M = N_1 \frac{\Phi_{21}}{I_2} \qquad (3.26)$$

If a fraction, k, of the flux leaving one coil links with the other coil, then

$$k = \frac{\Phi_{12}}{\Phi_1} = \frac{\Phi_{21}}{\Phi_2}$$

where $0 \leqslant k \leqslant 1.0$. Hence

$$\Phi_{12} = k\Phi_1 \quad \text{and} \quad \Phi_{21} = k\Phi_2$$

If the coils are wound on a constant-reluctance core, then multiplying equations 3.24 and 3.26 gives

$$M^2 = \left(N_2 \frac{\Phi_{12}}{I_1}\right) \times \left(N_1 \frac{\Phi_{21}}{I_2}\right) = \left(N_2 \frac{k\Phi_1}{I_1}\right) \times \left(N_1 \frac{k\Phi_2}{I_2}\right)$$

$$= k_2 \left(\frac{N_1 \Phi_1}{I_1}\right) \times \left(\frac{N_2 \Phi_2}{I_2}\right)$$

But $L_1 = N_1 \Phi_1/I_1$ and $L_2 = N_2 \Phi_2/I_2$; hence

$$M^2 = k^2 L_1 L_2$$

or

$$k = \frac{M}{\sqrt{(L_1 L_2)}} \qquad (3.27)$$

Example 3.11

The mutual inductance between two coils is 0.2 H. If the current in one winding increases from 100 mA to 600 mA in 5 ms, (a) determine the average value of e.m.f. induced in the secondary winding during this period of time, and (b) if the secondary is wound with 500 turns, calculate the change of flux linking with the secondary winding.

Solution

(a) $\mathrm{d}i = 600 - 100 = 500$ mA

   $\mathrm{d}t = 5$ ms

Average value of induced e.m.f. $= M\, di/dt$

$$= 0.2 \times 500 \times 10^{-3}/(5 \times 10^{-3})$$

$$= 20\ \text{V}$$

(b) From equation 3.23

$$d\Phi = M\, di_1/N_2$$

hence

flux change $= 0.2 \times 500 \times 10^{-3}/500 = 0.2 \times 10^{-3}$ Wb $= 0.2$ mWb

3.20   The Dot Notation

The dot notation is a convenient method of indicating the relative polarities of mutually induced e.m.f.s in magnetically coupled coils. In this notation a dot is placed arbitrarily at one end of the primary winding, and a second dot is then placed at the end of the magnetically coupled coil *which has the same instantaneous polarity as the 'dotted' end of the primary winding.* Referring to figure 3.17, if a dot is placed at the right-hand end of coil A, then we must also place a dot at the right-hand end of coil B since this has the same instantaneous polarity as the dotted end of coil A. Thus the coils in figure 3.17 can be represented by the circuit in figure 3.18a. The effect of reversing the direction of one of the windings is illustrated in figure 3.18b. Here the polarity of the induced e.m.f. is reversed, and the left-hand end of the secondary coil is instantaneously positive when the right-hand end of the primary is connected to the positive pole of the supply.

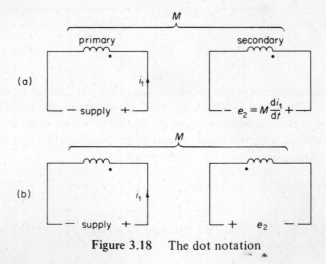

Figure 3.18   The dot notation

3.21 Series-connected Magnetically Coupled Coils

When two coils wound on the same former are electrically connected together, then the flux that links the coils will modify the value of the induced e.m.f. in the coils. In the absence of mutual coupling between the two coils of inductance L_1 and L_2, their inductance when they are connected in series is $L_1 + L_2$; the effect of mutual coupling is to alter the net value of the inductance.

Let us consider the arrangement in figure 3.19a in which the fluxes produced by the coils act in the same direction (the *series-aiding* connection). Each coil will have both self- and mutually induced e.m.f.s in them, and for coil L_1 these are

$$\begin{matrix} \text{self-induced e.m.f. due} \\ \text{to the current in } L_1 \end{matrix} = L_1 \frac{di}{dt}$$

$$\begin{matrix} \text{mutually induced e.m.f. due} \\ \text{to the current in } L_2 \end{matrix} = M \frac{di}{dt} \tag{3.28}$$

and for coil L_2 are

$$\begin{matrix} \text{self-induced e.m.f. due} \\ \text{to the current in } L_2 \end{matrix} = L_2 \frac{di}{dt}$$

$$\begin{matrix} \text{mutually induced e.m.f. due} \\ \text{to the current in } L_1 \end{matrix} = M \frac{di}{dt} \tag{3.29}$$

Therefore

$$\text{total induced e.m.f.} = (L_1 + L_2 + 2M) \frac{di}{dt} \tag{3.30}$$

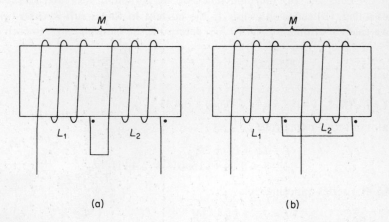

(a)   (b)

Figure 3.19   Coils connected in series so that the flux is (a) aiding and (b) opposing

If the equivalent inductance of the two coils in the series-aiding connection is L, then the induced e.m.f. in the circuit is

$$\text{total induced e.m.f.} = L \frac{di}{dt} \tag{3.31}$$

Comparing equations 3.30 and 3.31, we wee that the *equivalent inductance of two series-aiding coils* is

$$L = L_1 + L_2 + 2M \tag{3.32}$$

In the case of two coils connected in *series opposition*, figure 3.19b, the mutual flux induces an e.m.f. in opposition to the self-induced e.m.f. in the coils, and the polarities associated with the mutually induced e.m.f.s (equations 3.28 and 3.29) are negative. Hence the total induced e.m.f. in the circuit is

$$(L_1 + L_2 - 2M) \frac{di}{dt} \quad \text{volts}$$

to give an *equivalent for series-opposing coils* of

$$L = L_1 + L_2 - 2M \tag{3.33}$$

Example 3.12

Two coils are wound on a common magnetic circuit and have inductances of 1 H and 0.64 H. If the coupling coefficient is 0.5, determine (i) the mutual inductance between the coils and (ii) the inductance of the circuit if they are conected in (a) series-aiding, (b) series-opposing. If the current in the circuit in cases (a) and (b) above changes at the rate of 100 A/s, determine the induced e.m.f. in each case.

Solution

(i) $M = k\sqrt{(L_1 L_2)} = 0.5 \times \sqrt{(1 \times 0.64)} = 0.4\ \text{H}$

(ii) (a) For series-aiding coils

$$L = L_1 + L_2 + 2M$$
$$= 1 + 0.64 + (2 \times 0.4) = 2.44\ \text{H}$$

(b) For series-opposing coils

$$L = L_1 + L_2 - 2M$$
$$= 1 + 0.64 - (2 \times 0.4) = 0.84\ \text{H}$$

When the current changes at the rate of 100 A/s, the induced e.m.f. for

(a) series-aiding is

$$Ldi/dt = 2.44 \times 100 = 244 \text{ V}$$

(b) series-opposing is

$$L\, di/dt = 0.84 \times 100 = 84 \text{ V}$$

3.22 Energy Stored in a Magnetic Field

If the current flowing in an inductive circuit of inductance L increases at a uniform rate from zero to I amperes in a time of t seconds, then the *average* circuit current is $I/2$ amperes, and the *average* value of induced e.m.f. is $L \times$ rate of change of current or LI/t volts. Hence the *average* energy consumed by the inductive circuit is

$$W = EIt = \frac{LI}{t} \times \frac{I}{2} \times t = \frac{1}{2}LI^2 \quad \text{joules} \tag{3.34}$$

In the general case, the current rises at a non-linear rate and the instantaneous value of induced e.m.f. is $e = L\, di/dt$. The energy absorbed in time interval dt is

$$w = ei\, dt = L\frac{di}{dt} \times i \times dt = Li\, di \quad \text{joules}$$

The total amount of energy consumed during the time that the current changes from zero to I is

$$W = \int_0^I Li\, di = L\left[\frac{1}{2}i^2\right]_0^I = \frac{1}{2}LI^2 \quad \text{joules}$$

Example 3.13

A coil of resistance 20 Ω is connected to a 40-V d.c. supply. Calculate the steady value of current in the circuit. If the inductance of the coil is 5 H, determine the energy stored in the magnetic circuit.

Solution

During the time that the energy is being stored in the magnetic field, the current increases from zero to its final value. The rate of rise of current is determined by several factors including the supply voltage and the inductance and resistance of the coil (see also chapter 10). The final value of current is reached when the self-induced e.m.f. has fallen to zero, that is, when $di/dt = 0$, and the final current is limited by the resistance of the circuit. That is

$$\text{final value of current} = I = E/R = 40/20 = 2 \text{ A}$$

and the energy stored in the magnetic field is

$$W = \frac{1}{2}LI^2 = \frac{1}{2} \times 5 \times 2^2 = 10 \text{ J}$$

Summary of essential formulae and data

E.M.F. induced in a coil: $E = N \, d\Phi/dt$ volts

Flux density: $B = \Phi/A$ tesla

E.M.F. induced in a conductor: $E = Blv$ volts

Magnetic fringing: leakage coefficient $= \dfrac{\text{total magnetic flux}}{\text{useful magnetic flux}}$

Force on a conductor: $F = BIl$ newtons

Magnetomotive force: $F = IN$ ampere turns

Magnetising force: $H = F/l = IN/l$ ampere turns/metre

Absolute permeability: $\mu = B/H = \mu_r\mu_0$ henrys/metre
   permeability of free space $= \mu_0 = 4\pi \times 10^{-7}$ henrys/metre

Hysteresis loss: $P_h \propto f(B_{max})^n$, where n lies between about 1.6 and 2.2 and is typically 1.7

Reluctance: $S = F/\Phi = l/\mu_r\mu_0 \, A$ ampere turns/Wb

Permeance: $\Lambda = 1/S$ webers/ampere turn

E.M.F. of self inductance: $e = L \, di/dt$ volts

Inductance: $L = N \, d\Phi/di = \mu N^2 A/l = N^2/S$ henrys

E.M.F. of mutual inductance: $e_2 = M \, di_1/dt = N_2 \, d\Phi_1/dt$ volts

Mutual inductance: $M = N_2 \, d\Phi_1/di_1$

Coupling coefficient: $k = M/\sqrt{(L_1 L_2)}$

Inductance of two series-aiding coils: $L = L_1 + L_2 + 2M$

Inductance of two series-opposing coils: $L = L_1 + L_2 - 2M$

Energy stored in a magnetic field: $W = \dfrac{1}{2}LI^2$ joules

PROBLEMS

In the following, the permeability of free space (μ_0) has the value $4\pi \times 10^{-7}$ H/m.

3.1 A straight conductor of active length 50 cm cuts a magnetic field perpendicular to the line of action of the field. If the flux density is 0.2 T and the velocity of the conductor is 8 m/s, calculate the e.m.f. induced in the conductor.
[0.8 V]

3.2 If the conductor in problem 3.1 moves at an angle of (a) $60°$ and (b) 0.5 rad to the line of action of the magnetic field, determine the e.m.f. induced in the conductor.
[(a) 0.693 V; (b) 0.384 V]

3.3 Calculate the e.m.f. generated between the wing-tips of an aircraft which is flying horizontally at a velocity of 650 km/h if the vertical component of the earth's magnetic field is 40 μT. The wing span of the aircraft is 46 m.
[0.332 V]

3.4 A 250-turn coil is rotated in a uniform magnetic field of flux density 0.1 T. If the active length of each conductor is 0.25 m and the linear velocity of the conductors is 8 m/s, calculate the maximum value of the e.m.f. induced in the coil.
[100 V]

3.5 A straight conductor carries a current of 50 A, the length of the conductor being 20 cm. If the conductor is at right-angles to a magnetic field of flux density of 0.5 T, calculate the force acting on the conductor.
[5 N]

3.6 A galvanometer coil carries a current of 10 mA and has 25 turns of wire. If the active length of each side of the coil is 20 mm and it moves at right-angles to a uniform magnetic field of flux density 0.12 T, calculate the torque produced by the coil if its radius is 20 mm.
[0.48 μN m]

3.7 A mild steel ring is wound with 1500 turns of wire carrying a current of 2 A. If the resulting magnetising force is 2000 A/m, calculate the mean diameter of the ring.
[0.477 m]

3.8 A magnetic circuit has a mean length of 475 mm and a uniform cross-sectional area of 5 cm^2. If the magnetic circuit is uniformly wound with 189 turns of wire

which carry a current of 1 A, and the flux produced is 0.2 mWb, calculate the flux density in the core and also the relative permeability of the iron circuit.
[0.4 T; 795.8]

3.9 A magnetic circuit consists of three parts X, Y and Z; parts X and Y are of iron and Z is an air gap. Part X has a mean length of 40 cm and a cross-sectional area of 3 cm^2, part Y has a mean length of 15 cm and cross-sectional area of 3.5 cm^2, and part Z is an air gap of length 0.1 cm and is 3 cm^2 in area.

(a) If magnetic leakage can be neglected, determine the value of the current which must flow in a coil of 1000 turns of wire wound uniformly around the iron circuit if the flux density in the air gap is to be 1.0 T. The characteristic of the magnetic circuit is given in table 3.2. (b) If the magnetic circuit has a leakage co-efficient of 1.1 and the flux density in the air gap is to remain at 1.0 T, calculate the value of the current in the coil.

Table 3.2

B (T)	0.79	0.925	1.0	1.09	1.11
H (A/m)	500	700	900	1400	1600

[(a) 1.24 A; (b) 1.51 A]

3.10 A 680-turn coil is wound on the centre limb of the cast steel magnetic circuit in figure 3.20. If magnetic leakage can be neglected, determine the current required in the coil to produce a flux of 1.6 mWb in the air gap. The details of the *B–H* curve for the magnetic material are given in table 3.3.

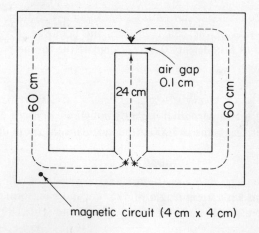

magnetic circuit (4 cm x 4 cm)

Figure 3.20

Table 3.3

B (T)	0.4	0.6	0.9	1.1	1.9
H (A/m)	470	590	800	1040	1200

[1.95 A]

3.11 When a current of 5 A flows in an air-cored coil of 1500 turns it produces a flux of 5 μWb. Calculate the inductance of the coil.
[1.5 mH]

3.12 A coil of 4000 turns is wound uniformly on a wooden ring of mean diameter 0.3 m and of cross-sectional area 5 cm^2. Determine the inductance of the coil.
[10.67 mH]

3.13 Coil A and coil B are wound on a common magnetic circuit. If the respective self-inductances are 0.16 H and 1.0 H, calculate the mutual inductance between the coils if the coils are coupled with a coupling coefficient of (a) $k = 1$, (b) $k = 0.1$.
[(a) 0.4 H; (b) 0.04 H]

3.14 When the current in coil P changes from +4 A to −4 A, the flux linkages produced in coil Q amount to 24 mWb-turns. Determine the mutual inductance existing between the two coils. Calculate also the self-inductance of coil Q if the coefficient of magnetic coupling is 0.5 and the self-inductance of coil P is 6 mH.
[3 mH; 6 mH]

3.15 Two coils X and Y are connected in series with one another, coil X having a self-inductance of 10 mH; the effective inductance of the combination is found to be 22 mH. When the connections to coil Y are reversed, the effective inductance is 18 mH. Determine (a) the self-inductance of coil Y, (b) the mutual inductance between the coils.
[(a) 10 mH; (b) 1 mH]

3.16 A coil is wound on a non-magnetic ring of length 1.0 m and of cross-sectional area 2.5 cm^2. When a certain current is passed through the coil, the flux density is found to be 0.06 T. Determine the energy stored in the magnetic circuit.
[0.358 J]

3.17 Determine the energy stored in the magnetic circuit of a coil of self-inductance 2.5 H when it carries a current of 2.4 A.
[7.2 J]

4 Electrostatics

4.1 Insulating Materials and Electric Charge

The application of an electrical potential between two metal *plates* or *electrodes* causes an *electric field* to be established in the insulation or *dielectric* between the electrodes, in the manner of figure 4.1a. The electric field is represented by imaginary lines, each of which shows the path that would be traced out by the movement of a free electric charge. The *direction of the electric field* at any point is given by the direction of the force experienced by a unit positive charge placed at that point.

The two electrodes together with the dielectric form a *capacitor*; the circuit symbols for both *fixed capacitors* and *variable capacitors* are shown in figure 4.1b. The *capacitance* of a variable capacitor can be adjusted by altering the physical parameters of the capacitor. The capacitance, symbol C, of a capacitor is its ability to store an electrical charge. The mechanism of charge storage depends on the chemical structure of the dielectric. The molecular structure of dielectric materials used in commercial capacitors falls into one of two categories, namely *polar molecule* (or *dipole*) and *non-polar molecule*.

In materials with the polar molecular structure, the centre of gravity of the electrons does not coincide with that of the atomic nuclei. Consequently the polar molecule may be regarded as the electrostatic equivalent of a bar magnet. When the capacitor is uncharged, the axes of the molecules in the dielectric assume random directions and the net electrical potential between opposite faces of the material is zero. The application of an electric field to the dielectric causes the molecules to pivot about their centre of mass (that is, about the positively charged atomic nuclei), the 'lighter end' being attracted towards the positive electrode; this causes the dielectric to be in a state of electric strain. When the voltage applied to the dielectric is removed the molecules remain strained, and in this way energy is stored in the capacitor.

The orbits of electrons in non-polar molecules are capable of elastic strain, and when non-polar materials are placed in an electric field they become polar and are capable of storing charge in the same way as polar materials.

The majority of insulating materials are of the polar type and include cellulose, phenol—formaldehyde resins and PVC. Non-polar types include polystyrene, polythene and transformer oil.

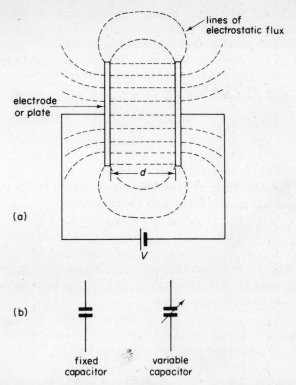

Figure 4.1 (a) The electric field produced by a capacitor and (b) circuit symbols used to represent the capacitor

The *electric field intensity* (also described as the *electric stress*, the *electric field strength* or the *potential gradient*), symbol E, experienced by the dielectric is given by

$$E = V/d \quad \text{volts per metre} \tag{4.1}$$

where V and d are defined in figure 4.1a. The maximum electric stress a material can sustain without breaking down is known as its *electric strength*, and the voltage which has to be applied to cause electrical breakdown is known as the *breakdown voltage*.

4.2 The Relationship between Charge, Capacitance and Applied Voltage

Experiments show that the electrical charge Q stored by a capacitor is given by the relationship

$$Q = CV \quad \text{coulombs} \tag{4.2}$$

where C is the capacitance in *farads*, symbol F, and V is the voltage between the plates of the capacitor. The charge in coulombs represents the displacement in the external electrical circuit of a given number of electrons, and is equal to the *quantity* of electricity that has passed through the circuits.

The farad is an inconveniently large unit, and subunits used in practice are the microfarad (μF), the nanofarad (nF) and the picofarad (pF), where

$$1\mu F = 10^{-6} \text{ F}$$

$$1 \text{ nF} = 10^{-9} \text{ F}$$

$$1 \text{ pF} = 10^{-12} \text{ F}$$

The farad is defined from equation 4.2 as follows.

The farad is the capacitance of a capacitor which stores a charge of one coulomb when a p.d. of one volt appears between its electrodes.

Example 4.1

A potential of 400 V is maintained between the electrodes of a capacitor of 0.2 nF capacitance. Calculate (a) the stored charge, and (b) the potential gradient in the dielectric given that its thickness is 2 mm.

Solution

(a)

$$Q = CV = 0.2 \times 10^{-9} \times 400 = 80 \times 10^{-9} \text{ C}$$

$$= 80 \text{ nC} = 0.08 \ \mu C$$

(b)

$$E = V/d = 400/2 \times 10^{-3} = 200 \times 10^{3} \text{ V/m}$$

$$= 200 \text{ kV/m}$$

4.3 Capacitors in Parallel

The parallel-connected capacitors in figure 4.2a both support voltage V between their terminals. The charge stored by C_1 is

$$Q_1 = C_1 V$$

and the charge stored by C_2 is

$$Q_2 = C_2 V$$

If the parallel combination were replaced by a single capacitor C, as shown in figure 4.2b, its value being such that its stored charge is $Q_1 + Q_2$, then

$$CV = Q_1 + Q_2 = C_1 V + C_2 V = V(C_1 + C_2)$$

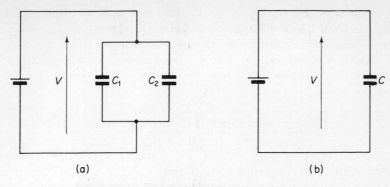

Figure 4.2   Parallel-connected capacitors

or

$$C = C_1 + C_2 \qquad (4.3)$$

Hence, *the effective capacitance of two capacitors in parallel is the sum of their respective values.* This argument can be extended to deal with *n* capacitors in parallel, to give an equivalent capacitance of

$$C = C_1 + C_2 + \ldots + C_n \qquad (4.4)$$

Note The effective capacitance of parallel-connected capacitors is *always greater than* the largest individual value of capacitance.

Example 4.2

Calculate the equivalent capacitance of three parallel-connected capacitors of values 0.5 μF, 2000 pF and 200 nF.

Solution

First the values need to be changed to equivalent unit sizes.

$$C_1 = 500 \text{ nF} \quad C_2 = 2 \text{ nF} \quad C_3 = 200 \text{ nF}$$
$$C = C_1 + C_2 + C_3 = 500 + 2 + 200 = 702 \text{ nF or } 0.702 \text{ } \mu\text{F}$$

4.4 Series-connected Capacitors

When switch S is closed in figure 4.3a the charging current flows through both capacitors for the same length of time. Since $Q = It$, the charge stored by each capacitor is Q coulombs. If V_1 and V_2 are the p.d.s across C_1 and C_2, respectively, then

$$Q = C_1 V_1 = C_2 V_2$$

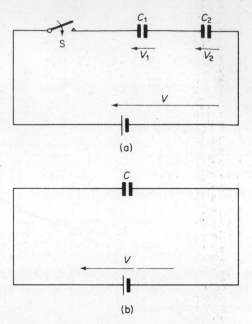

Figure 4.3   Series-connected capacitors

or

$$V_1 = Q/C_1 \quad \text{and} \quad V_2 = Q/C_2$$

If we replace the two series-connected capacitors in figure 4.3a by the single capacitor in figure 4.3b such that it stores the same charge Q coulombs when V volts are applied to it, then

$$Q = CV \quad \text{or} \quad V = Q/C$$

Comparing the two circuits we see that

$$V = V_1 + V_2$$

or

$$\frac{Q}{C} = \frac{Q}{C_1} + \frac{Q}{C_2}$$

therefore

$$\frac{1}{C} = \frac{1}{C_1} + \frac{1}{C_2}$$

That is, *the reciprocal of the equivalent capacitance of two series-connected capacitors is the sum of the reciprocals of their respective capacitances.*

Rearranging yields

$$C = \frac{C_1 C_2}{C_1 + C_2} \tag{4.5}$$

In the general case of n capacitors connected in series, the reciprocal of the equivalent capacitance is

$$\frac{1}{C} = \frac{1}{C_1} + \frac{1}{C_2} + \dots + \frac{1}{C_n} \tag{4.6}$$

Note The equivalent capacitance of series-connected capacitors is *always less than* the value of the smallest capacitance in the circuit.

Example 4.3

Calculate the equivalent capacitance of three series-connected capacitors of values 0.5 μF, 2000 pF and 200 nF.

Solution

$$C_1 = 0.5\ \mu\text{F} \quad C_2 = 0.002\ \mu\text{F} \quad C_3 = 0.2\ \mu\text{F}$$

$$\frac{1}{C} = \frac{1}{0.5} + \frac{1}{0.002} + \frac{1}{0.2} = 2 + 500 + 5 = 507\ (\mu\text{F})^{-1}$$

or

$$C = 1/507 = 0.00197\ \mu\text{F} \quad \text{or} \quad 1970\ \text{pF}.$$

4.5 Voltage Distribution between Series-connected Capacitors

From the work in section 4.4 on series circuits, the same charge is stored by each capacitor. For the circuits in figures 4.3a and b

$$Q = CV = C_1 V_1 = C_2 V_2$$

therefore

$$CV = C_1 V_1$$

or

$$V_1 = \frac{C}{C_1} V = \frac{[C_1 C_2 / (C_1 + C_2)]}{C_1} V = \frac{C_2}{C_1 + C_2} V \tag{4.7}$$

It may also be shown that

$$V_2 = \frac{C_1}{C_1 + C_2} V \qquad (4.8)$$

In the general case of n capacitors connected in series, the voltage V_g across the gth capacitor is

$$V_g = CV/C_g \qquad (4.9)$$

where C is the equivalent capacitance of the series circuit and C_g is the capacitance of the gth capacitor.

The above equations have practical relevance in the calculation of the static voltage distribution across insulator strings and other devices which have self-capacitance—for example, semiconductor devices such as thyristors and triacs in their non-conducting states.

Example 4.4

Capacitors of 1, 10 and 100 μF are connected in series to a 222-V d.c. supply. Calculate the voltage across each capacitor.

Solution

$$C_1 = 1\ \mu\text{F}, \quad C_2 = 10\ \mu\text{F}, \quad C_3 = 100\ \mu\text{F}$$

$$C = 1/\left(\frac{1}{C_1} + \frac{1}{C_2} + \frac{1}{C_3}\right) = 1/\left(\frac{1}{1} + \frac{1}{10} + \frac{1}{100}\right) = 0.9\ \mu\text{F}$$

From equation 4.9

$$V_1 = CV/C_1 = 0.9 \times 10^{-6} \times 222/1 \times 10^{-6} = 200\ \text{V}$$
$$V_2 = CV/C_2 = 0.9 \times 10^{-6} \times 222/10 \times 10^{-6} = 20\ \text{V}$$
$$V_3 = CV/C_3 = 0.9 \times 10^{-6} \times 222/100 \times 10^{-6} = 2\ \text{V}$$

Note From the above we see that the largest value of capacitance supports the smallest value of voltage and vice versa.

4.6 Series–Parallel Capacitor Combinations

The effective capacitance of the circuit in figure 4.4a is calculated by sub-dividing it into two sections, namely a series combination C_S and a parallel combination C_P, where

$$C_S = C_1 C_2/(C_1 + C_2)$$

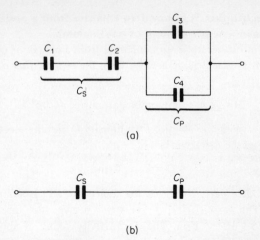

(a)

(b)

Figure 4.4   A series–parallel combination of capacitors

and

$$C_P = C_3 + C_4$$

The circuit is reduced to that in figure 4.4b whose effective capacitance C is

$$C = C_S C_P / (C_S + C_P)$$

Example 4.5

In a circuit similar to that in figure 4.4a, $C_1 = 0.01~\mu F$, $C_2 = 0.1~\mu F$, $C_3 = 0.2~\mu F$, and $C_4 = 1~\mu F$. Calculate the effective capacitance of the circuit.

Solution

$$C_S = 0.01 \times 0.1/(0.01 + 0.1) = 0.0091~\mu F$$

$$C_P = 0.2 + 1 = 1.2~\mu F$$

and

$$C = C_S C_P / (C_S + C_P) = 0.0091 \times 1.2/(0.0091 + 1.2)$$

$$= 0.009~\mu F$$

4.7 Electric Flux Density and Permittivity

It was stated in section 4.1 that a free charge in an electric field experiences a force, and if allowed to move in the field would trace out a curve. This leads to the concept of *lines of electric flux*.

One line of electric flux is assumed to emanate from a positive charge of one coulomb, and to enter a negative charge of one coulomb.

Hence, Q lines of electrostatic flux emanate from a charge of Q coulombs. If the flux passes through a dielectric of area A, then the *electric flux density*, symbol D, in the dielectric is

$$D = Q/A \quad \text{coulombs per metre}^2 \ (\text{C/m}^2) \tag{4.10}$$

The relationship between the electric flux density D and the electric field intensity E (see section 4.1) in the dielectric is

$$D = E\epsilon \quad \text{C/m}^2 \tag{4.11}$$

where ϵ is the *absolute permittivity* of the dielectric, and has dimensions of farads per metre (F/m). The *permittivity of free space* is given the special symbol ϵ_0 and has the value 8.85×10^{-12} F/m. The permittivity of air is about 0.06 per cent greater than that of free space and, for all practical purposes, the permittivities of free space and air can be taken to be equal to one another.

When the space between the plates of a capacitor is filled with an insulating medium such as glass or mica then, for a given voltage between the plates, the electric flux density (and capacitance) is increased. The ratio of the increase when compared with the case when the dielectric is air is indicated by the *relative permittivity*, ϵ_r, of the material. The relative permittivity is simply a number and is therefore dimensionless. Typical values of ϵ_r are given in table 4.1.

Table 4.1

Material	Relative permittivity
Air	1.0006
Bakelite	4.5–5.5
Glass	5–10
Mica	3--7
Paper (dry)	2–2.5
Rubber	2–3.5

The absolute permittivity is related to the permittivity of free space as follows

$$\epsilon = \epsilon_0 \, \epsilon_r \tag{4.12}$$

4.8 Capacitance of a Parallel-plate Capacitor

The simplest form of capacitor is the parallel-plate capacitor shown in figure 4.5, the plates of area A m^2 being d metres apart. In the following we assume that all lines of electric flux pass directly from one plate to the other, and that we can

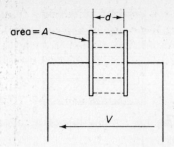

Figure 4.5 Capacitance of a parallel-plate capacitor

neglect the effects of *electrostatic fringing*. The electric field strength in the dielectric is

$$E = V/d \quad \text{V/m}$$

and the electrostatic flux density in the dielectric is

$$D = Q/A \quad \text{C/m}^2$$

where Q is the electric charge stored by the capacitor. Now, from equation 4.11

$$\epsilon = \frac{D}{E} = \frac{Q}{A} \times \frac{d}{V} = \frac{Q}{V} \times \frac{d}{A}$$

But, since $Q = CV$, then

$$\epsilon = C \times \frac{d}{A} \tag{4.13}$$

or

$$C = \frac{\epsilon A}{d} = \frac{\epsilon_0 \epsilon_r A}{d} \tag{4.14}$$

In the case of a **multiple-plate capacitor** of the kind in figure 4.6, the five plates are separated by four dielectrics (since only one side of each of the outer plates forms part of the capacitor). In the general case of an *n*-plate capacitor there are $(n - 1)$ dielectrics, and the *effective area* of an equivalent two-plate capacitor is $(n - 1)A$, where A is the area of one of the plates in figure 4.6. The capacitance of the multi-plate capacitor is therefore

$$C = \frac{(n - 1)\epsilon_0 \epsilon_r A}{d} \quad \text{farads} \tag{4.15}$$

Example 4.6

A parallel-plate capacitor consists of two metal plates each of area 500 cm^2 separated by a dielectric of thickness 1.5 mm whose relative permittivity is 5. When

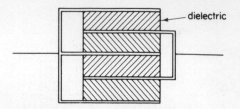

Figure 4.6   A multiple-plate capacitor

the voltage between the plates is 400 V, determine (a) the capacitance of the
capacitor in pF, (b) the charge stored in μC, (c) the electric field intensity in kV/m,
and (d) the electric flux density in $\mu C/m^2$.

Solution

(a) $C = \epsilon_0 \, \epsilon_r \, A/d = 8.85 \times 10^{-12} \times 5 \times 500 \times (10^{-2})^2 / 1.5 \times 10^{-3}$

        $= 147.5 \times 10^{-11} \text{ F} = 1.475 \text{ nF}$

(b) $Q = CV = 147.5 \times 10^{-11} \times 400 \text{ C} = 0.59 \ \mu C$

(c) $E = V/d = 400/1.5 \times 10^{-3} \text{ V/m} = 266.7 \text{ kV/m}$

(d) $D = Q/A = 0.59/500 \times (10^{-2})^2 \ \mu C/m^2 = 11.8 \ \mu C/m^2$

4.9 Capacitors with Composite Dielectrics

In several practical forms of insulator the dielectric is constructed from several
different types of insulating material. One form of capacitor with a composite
dielectric is shown in figure 4.7a, in which dielectrics X and Y are of different
materials. For practical purposes we may regard the dielectrics as forming two
series-connected capacitors, shown in figure 4.7b, whose effective capacitance C is

$$C = C_1 C_2 / (C_1 + C_2)$$

where

$$C_1 = \epsilon_0 \, \epsilon_{r1} \, A/d_1 \quad \text{and} \quad C_2 = \epsilon_0 \epsilon_{r2} A/d_2$$

in which d_1, d_2 and A are defined in figure 4.7a, and ϵ_{r1} and ϵ_{r2} are the relative
permittivities of materials X and Y, respectively. From the work in section 4.5, the
voltage across dielectric X is

$$V_1 = CV/C_1$$

and the voltage V_2 across dielectric Y is

$$V_2 = CV/C_2$$

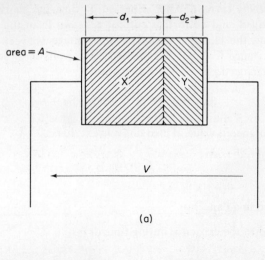

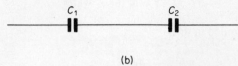

(b)

Figure 4.7   A capacitor with a composite dielectric

The electric field strengths (or potential gradients) in the two materials are

Dielectric X $E_1 = V_1/d_1 = CV/C_1 d_1$

Dielectric Y $E_2 = V_2/d_2 = CV/C_2 d_2$

Hence

$$\frac{E_1}{E_2} = \frac{C_2 d_2}{C_1 d_1}$$

Applying the results of equation 4.14 to the above equation yields

$$\frac{E_1}{E_2} = \frac{\epsilon_{r2}}{\epsilon_{r1}} \qquad (4.16)$$

Thus if $\epsilon_{r1} = 1.5$ and $\epsilon_{r2} = 4.5$, then $E_1 = 3 E_2$.

4.10 Capacitor Charging Current

When an uncharged capacitor is connected to an electrical supply it begins to store energy. During the time that it absorbs energy from the supply, it draws a current known as the *charging current*. After a period of time has elapsed the capacitor becomes fully charged, and no further current is drawn from the supply (see also chapter 10). During the charging period, if the voltage between the plates of a capacitor of capacitance C farads is increased by dv volts in dt seconds, then the increase in stored charge dq is

$$dq = i\, dt = C\, dv \quad \text{coulombs}$$

where i is the instantaneous value of charging current; hence

$$i = C\, dv/dt$$

4.11 Energy Stored in a Capacitor

The power supplied to the capacitor during time dt is

$$vi = v\, C\, dv/dt \quad \text{watts}$$

and the energy w supplied in dt is

$$w = vi\, dt = vC\frac{dv}{dt}\, dt = vC\, dv \quad \text{joules}$$

The energy supplied to the capacitor when the p.d. is increased from zero volts to V volts is

$$W = \int_0^V vC\, dv = \frac{1}{2}\, C[v^2]_0^V = \frac{1}{2}CV^2 \quad \text{joules} \qquad (4.17)$$

Example 4.7

A capacitor of $100\,\mu\text{F}$ has a p.d. of $10\,\text{V}$ between its plates. Calculate the amount of energy stored by the capacitor.

Solution

$$\text{Energy stored} = W = \frac{1}{2}CV^2 = \frac{1}{2} \times 100 \times 10^{-6} \times 10^2 = 0.005\,\text{J}$$

4.12 Summary of Electric, Electromagnetic and Electrostatic Quantities

Analogous quantities in electric, electromagnetic and electrostatic circuits referred to in this and earlier chapters are collected together in table 4.2.

Table 4.2

Electric		Electromagnetic		Electrostatic	
Current	I	Magnetic flux	Φ	Electric flux	Q
Voltage	V	m.m.f.	F	Voltage	V
Resistance	R	Reluctance	S		
Conductance	G			Capacitance	C
Current density		Magnetic flux density	B	Electric flux density	D
Potential gradient		Magnetising force	H	Electric field intensity	E
Conductivity		Permeability	μ	Permittivity	ϵ

Summary of essential formulae and data

Electric field intensity: $E = V/d$ volts per metre

Electric charge: $Q = CV$ coulombs

Capacitors in parallel: $C = C_1 + C_2 + \dots$ farads

Capacitors in series: $1/C = 1/C_1 + 1/C_2 + \dots$ (farads)$^{-1}$

Voltage distribution between two capacitors C_1 and C_2 in series:
  voltage across C_1 = applied voltage $\times C_2/(C_1 + C_2)$ volts

Electric flux density: $D = Q/A$

Absolute permittivity: $\epsilon = D/E = \epsilon_r \epsilon_0$ farads per metre
  permittivity of free space = $\epsilon_0 = 1/(4\pi \times 9 \times 10^9)$ farads per metre
    $= 8.85 \times 10^{-12}$ F/m

Capacitance of a multiple-plate parallel-plate capacitor: $C = (n-1)\epsilon A/d$ farads

Energy stored in a capacitor: $W = \dfrac{1}{2}CV^2$ joules

Current: $i = C \, dv/dt$ amperes

PROBLEMS

Note: $\mu_0 = 8.85 \times 10^{-12}$ F/m.

4.1 A d.c. potential of 500 V is applied between the electrodes of a parallel-plate capacitor, each plate having an area of 0.04 m^2. If the capacitance of the capacitor is 0.05 μF and the absolute permittivity of the dielectric is 2×10^{-11} F/m, determine (a) the electric flux density and (b) the electric field intensity in the dielectric.
[(a) 0.625 mC/m^2; (b) 31.25 MV/m]

4.2 Two capacitors having capacitances of 0.1 μF and 0.01 μF respectively are connected (a) in parallel, (b) in series with one another. Determine the equivalent capacitance in each case.
[(a) 0.11 μF; (b) 0.0091 μF]

4.3 Two 10-μF capacitors are connected in series with one another. A 4-μF capacitor is connected in parallel with the series combination. Determine the capacitance of the circuit.
[2.22 μF]

4.4 Two capacitors of capacitances 8 μF and 4 μF respectively are connected in parallel with one another. A third capacitor, C_3, is connected in series with the parallel combination to give an overall capacitance of 2 μF. Determine the capacitance of C_3.
[2.4 μF]

4.5 A parallel-plate capacitor has a capacitance of 500 pF. Determine the thickness of the dielectric if the relative permittivity of the dielectric is 8, and the area of each plate is 50 cm^2.
[0.71 mm]

4.6 A multiple-plate capacitor has 15 plates, and has a capacitance of 0.1 μF. If each dielectric is 0.5 mm thick and its relative permittivity is 2.5, calculate the area of each plate.
[0.168 m^2]

4.7 A parallel-plate capacitor has a dielectric consisting of a sheet of mica 1 mm thick (of relative permittivity 6) and a 6 mm thick layer of oil (of relative permittiity 2.2). If 500 V d.c. is applied between the plates, determine the p.d. across (a) the mica, (b) the oil.
[(a) 471.2 V; (b) 28.8 V]

4.8 A capacitor consists of two parallel plates separated by a sheet of insulating material 2.5 mm thick of relative permittivity 4.0. The separation between the plates is increased to allow a second dielectric of thickness 4.5 mm to be inserted. If the capacitance of the capacitor so formed is 40 per cent of that of the original capacitor, determine the relative permittivity of the second dielectric.
[4.8]

4.9 A parallel-plate capacitor is used as a displacement transducer, the dielectric of the capacitor being mechanically connected to the object whose displacement is to be measured. Show that if the dielectric, whose relative permittivity is 4, is displaced so that 75 per cent of the dielectric material remains between the capacitor plates, then the capacitance of the capacitor is reduced to 81.25 per cent of its maximum value. Neglect the effect of electrostatic fringing.

4.10 Two capacitors have capacitances of 0.1 μF and 0.5 μF, respectively. Determine the energy stored when they are connected (a) in series, (b) in parallel with one another to a 10-V d.c. supply.
[(a) 4.165 μJ; (b) 30 μJ]

4.11 Capacitors of 0.1 μF, 1 μF and 10 μF, respectively, are connected in series. If the energy stored in the 0.1-μF capacitor is 5 μJ, calculate the value of the supply voltage.
[11.1 V]

5 Alternating Voltage and Current

5.1 Generating an Alternating E.M.F.

Alternating voltage systems are almost universally used for the transmission of electrical power due to the ease not only with which the voltage can be generated, but also with which the magnitude of the supply voltage can be changed by transformer action (see also chapter 9).

A simple method of generating an alternating e.m.f. is shown in figure 5.1, in which a permanent magnet is rotated inside a coil of wire. The e.m.f.s induced in conductors A and B are additive and, under the conditions in the figure (applying Fleming's right-hand rule), terminal A′ is instantaneously negative with respect to terminal B′. A short time later when the magnet has turned through 180°, the induced e.m.f.s in the coil sides have both reversed and A′ is positive with respect to B′. Thus the current in the external circuit connected to terminals A′B′ pulsates or *alternates*. The alternating voltage may also be generated by rotating the coil inside a fixed magnetic field. Connections are made to the ends of the coil via rotating *slip rings* and *brushes*, shown in figure 5.2. The brushes in commercial machines are made of carbon.

The waveform of the e.m.f. induced in the coil depends not only on the construction of the coil but also on the magnetic field distribution. The shape of the e.m.f. wave generated by figure 5.1 would be similar to the flat-topped wave in figure 5.3a. This waveform is not particularly suited for the purposes of electrical transmission, the most suitable being the sinusoidal wave in figure 5.3b. The time for one complete cycle of the waveform to be generated is known as its *periodic time, T,* and the number of complete cycles generated in one second is

$$f = 1/T \quad \text{hertz (Hz)} \tag{5.1}$$

Commonly used power-supply frequencies are 50 Hz and 60 Hz, having periodic times of 20 ms and 16.67 ms, respectively.

The value of the generated e.m.f. varies between the extremes of $+E_m$ and $-E_m$; in the case of the sine wave, figure 5.3b, the instantaneous value of e.m.f. e varies sinusoidally according to the relationship

$$e = E_m \sin \theta \tag{5.2}$$

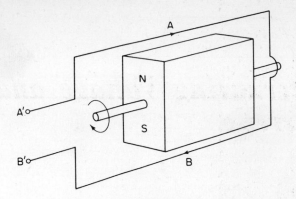

Figure 5.1 Principle of the alternator

where E_m is the maximum (peak) voltage and θ is the angle through which the magnet in figure 5.1 has turned from the horizontal.

5.2 The Average Value or Mean Value of an Alternating Waveform

The average value of any signal during a given time interval is

$$\text{average value} = \frac{\text{area under graph}}{\text{time interval}}$$

When evaluating the average value of an alternating wave, it is usual to do so *over one half-cycle of the wave* for the following reason. If we consider the waveforms in figures 5.3a and b, in each case the area under the positive half-wave is equal to that under the negative half-wave; that is, the *total area* under a complete cycle is zero. Hence *the average value of an alternating waveform taken over a complete cycle is zero*. Since the area under either half-cycle is finite, then the average value is taken to be that of a half-cycle.

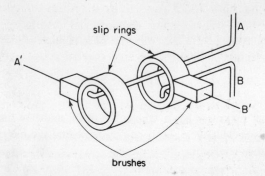

Figure 5.2 Slip-ring connections to a rotating loop of wire

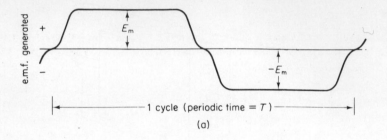

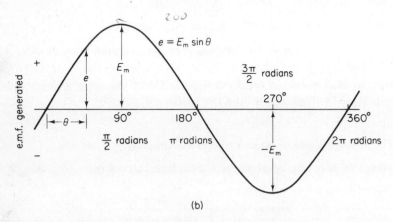

Figure 5.3   Alternating voltage waveforms

The average value of an alternating waveform can be computed in the manner shown in figure 5.4. In this case we are dealing with a current waveform, and if the values of n equidistant mid-ordinates are i_1, i_2, . . ., i_n, then the average value of the waveform over one half-cycle is

$$I_{av} = \frac{i_1 + i_2 + \ldots + i_n}{n} \qquad (5.3)$$

In the case of figure 5.4, $n = 6$. The average value of a voltage waveform is calculated using the same principle.

Example 5.1

The waveform of a current has a triangular shape and has the following values measured during one half-cycle, both half-cycles being symmetrical. Calculate the average value of the waveform.

Current (mA)	0	10	20	30	40	50	40	30	20	10	0
Time (ms)	0	10	20	30	40	50	60	70	80	90	100

Solution

Since the waveform of current changes linearly† between the intervals of time in the table, then the mid-ordinate values of current are 5, 15, 25, 35, 45, 45, 35, 25, 15, and 5 mA, respectively. Hence

$$I_{av} = \frac{5 + 15 + 25 + 35 + 45 + 45 + 35 + 25 + 15 + 5}{10}$$

$$= 25 \text{ mA}$$

5.3 Root-mean-square (R.M.S.) Value of an Alternating Waveform

The *effective value* of current in an a.c. circuit is computed in terms of its *heating effect*. Considering the waveform in figure 5.4, the instantaneous heating effect of current i_1 when flowing through a resistor of value R is $i_1^2 R$, and that due to i_2 is $i_2^2 R$, etc. Hence the *average heating effect* of the current waveform is

$$\frac{i_1^2 R + i_2^2 R + \ldots + i_n^2 R}{n}$$

If I is the *effective value* of alternating current that produces the same heating effect when flowing through R, then

$$I^2 R = \frac{i_1^2 R + i_2^2 R + \ldots + i_n^2 R}{n}$$

or

$$I = \sqrt{\left(\frac{i_1^2 + i_2^2 + \ldots + i_n^2}{n}\right)} \tag{5.4}$$

= square *root* of the *mean* of the sum of the *squares* of the current

= root-mean-square (r.m.s.) value of the current

The r.m.s. value of a voltage waveform is calculated using the above technique.

The r.m.s. value can be computed either over a complete cycle or over half a cycle since, when the instantaneous values of current (or voltage) are multiplied by themselves the results are always positive. Consequently, both half cycles of the (current)^2 waveform have positive values.

Example 5.2

Calculate the r.m.s. value of the current for the waveform in example 5.1.

†In most practical cases the waveform follows a smooth curve, and it is usually necessary to measure the mid-ordinates from the curve.

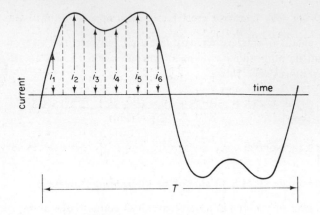

Figure 5.4   Determination of the average value of an alternating waveform

Solution

The instantaneous values of i and i^2 are as follows

Time (ms)	i (mA)	i^2 (mA)^2
5 and 95	5	25
15 and 85	15	225
25 and 75	25	625
35 and 65	35	1225
45 and 55	45	2025

The sum of the values of i^2 in the above table is 4125 mA^2, therefore the sum of the values of i^2 over one half-cycle is

$$2 \times 4125 = 8250 \text{ mA}^2$$

Hence

$$\text{r.m.s. value of current} = I = \sqrt{(8250/10)} = 28.72 \text{ mA}$$

Note The above value differs from the true solution by about 0.5 per cent due to the fact that only ten mid-ordinates were used. The accuracy of calculation is improved by increasing the number of mid-ordinates (see also section 5.5).

5.4 Form Factor and Peak Factor (Crest Factor)

It is the case that in all alternating waves with the exception of a rectangular wave (see below), the r.m.s. value is greater than its average value. The form factor of the wave, and also its peak factor or crest factor give an indication of the waveshape, the respective factors being defined below.

$$\text{Form factor} = \frac{\text{r.m.s. value of wave}}{\text{average value of wave}} = \frac{I}{I_{av}} \tag{5.5}$$

$$\frac{\text{Peak factor or}}{\text{crest factor}} = \frac{\text{peak value or maximum value}}{\text{r.m.s. value}} = \frac{I_m}{I} \tag{5.6}$$

In the case of a rectangular wave the average, r.m.s. and peak values are all equal to one another.

Example 5.3

Calculate the form factor and the peak factor of the triangular waveform in examples 5.1 and 5.2.

Solution

$$I_{av} = 25 \text{ mA}, I_m = 50 \text{ mA}, I = 28.72 \text{ mA}$$
$$\text{Form factor} = I/I_{av} = 28.72/25 = 1.15$$
$$\text{Peak factor} = I_m/I = 50/28.72 = 1.74$$

5.5 Average and R.M.S. Values of a Sinusoidal Waveform

To determine the average value of any waveform, it is first necessary to calculate the area enclosed by one half-cycle. Mathematically this is done by integrating the equation of the wave which is, in effect, the area computed using an infinite number of mid-ordinates. In the case of a sine wave, the equation for the instantaneous current, *i*, is

$$i = I_m \sin \theta$$

where I_m is the maximum value of current, and θ is the angle in radians from the instant of zero current (see also figure 5.3b for a sinusoidal voltage waveform). The area enclosed by the positive half-cycle of the wave is

$$\int_0^\pi i \, d\theta = \int_0^\pi I_m \sin \theta \, d\theta = -I_m [\cos \theta]_0^\pi = -I_m [-1 - 1]$$

$$= 2I_m \text{ ampere radians}$$

The average value of the current waveform taken over the half cycle is

$$I_{av} = \frac{\text{area under one half cycle}}{\pi \text{ (radians)}}$$

$$= 2I_m/\pi = 0.637 \, I_m \tag{5.7}$$

From equation 5.4, the r.m.s. current I is computed from the equation

$$I^2 = \frac{\text{total area under current}^2 - \text{angle graph}}{2\pi \text{ (radians)}}$$

$$= \frac{1}{2\pi} \int_0^{2\pi} i^2 \, d\theta = \frac{1}{2\pi} \int_0^{2\pi} I_m^2 \sin^2 \theta \, d\theta$$

$$= \frac{I_m^2}{2\pi} \int_0^{2\pi} \frac{1}{2}(1 - \cos 2\theta) \, d\theta = \frac{I_m^2}{4\pi} \left[\theta - \frac{1}{2}\cos 2\theta \right]_0^{2\pi}$$

$$= \frac{I_m^2}{2}$$

Hence

$$I = I_m/\sqrt{2} = 0.707 \, I_m \tag{5.8}$$

Similarly, for voltage waveforms

$$\text{average value} = V_{av} = 2V_m/\pi = 0.637 \, V_m \tag{5.9}$$

$$\text{r.m.s. value} = V = V_m/\sqrt{2} = 0.707 \, V_m \tag{5.10}$$

Consequently the form factor and the peak factor for sinusoidal waveforms are

$$\text{form factor} = I/I_{av} = 0.707 \, I_m/0.637 \, I_m = 1.11 \tag{5.11}$$

$$\text{peak factor} = I_m/I = I_m/0.707 \, I_m = 1.414 \tag{5.12}$$

Example 5.4

A sinusoidal voltage waveform of peak value 14.14 V is applied to a resistance of value 5 Ω. Calculate the r.m.s. value of voltage across the resistance, and the r.m.s. and peak values of circuit current.

Solution

From equation 5.10

$$\text{r.m.s. voltage} = V = 0.707 \, V_m = 0.707 \times 14.14 = 10 \text{ V}$$
$$\text{r.m.s. current} = I = V/R = 10/5 = 2 \text{ A}$$

From equation 5.8

$$I_m = \sqrt{2}I = 1.414 \times 2 = 2.828 \text{ A}$$

Note I_m can also be calculated from

$$I_m = V_m/R = 14.14/5 = 2.828 \text{ A}$$

5.6 Graphical Representation of Sinusoidal Quantities

If a line of length OA, see figure 5.5, rotates about point O at a constant velocity of ω radians per second in an anticlockwise direction, then the projection of point A on the vertical axis traces out the sinewave shown. The counterclockwise movement is a convenient direction of rotation, and is used throughout the book. The length of OA is equal to the maximum value of current. At the instant of time when the line is in position OA, the instantaneous value of current i_1 is given by the expression

$$i_1 = I_m \sin \theta_1$$

Some time later when the line has rotated to position OA', the instantaneous value of circuit current is

$$i_2 = I_m \sin \theta_2$$

At the completion of each revolution a new cycle is commenced. Since the line rotates through 2π radians during each revolution or cycle of events then, if the supply frequency is f Hz the angular frequency, ω, is

$$\omega = 2\pi f \quad \text{rad/s} \tag{5.13}$$

or

$$f = \omega/2\pi \quad \text{Hz} \tag{5.14}$$

Also, since

$$\theta = \text{angular velocity} \times \text{time} = \omega t \quad \text{radians}$$

then the expression for instantaneous current is

$$i = I_m \sin \theta = I_m \sin \omega t = I_m \sin 2\pi f t \tag{5.15}$$

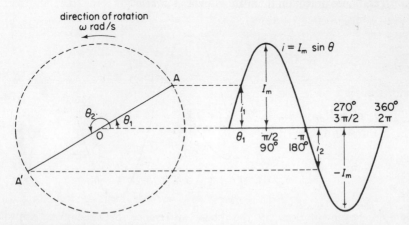

Figure 5.5 Graphical representation of an alternating waveform

5.7 Representation of Phase Angle Difference

In electrical engineering we are concerned with circuits in which quantities (both voltage and current) not only may differ in magnitude, but may also have a difference of *phase angle* between them. Two such waveforms are illustrated in figure 5.6. In the figure, current i_1 has zero value when OA lies on the horizontal axis as shown; at the same instant of time current i_2 has its maximum negative value. The angular difference, ϕ, between lines OA and OB, which trace out the sinewaves i_1 and i_2, is known as the *phase angle difference* between the waveforms. In the case shown it is $90°$ or $\pi/2$ radians.

As the lines OA and OB rotate in an anticlockwise direction, so i_2 passes through zero *after* i_1 has passed through zero. To indicate this fact we say that i_2 *lags* behind i_1 by angle ϕ. Alternatively, if we use i_2 as the reference waveform, we may say that i_1 *leads* i_2 by angle ϕ . In figure 5.6, the expression describing i_1 is

$$i_1 = I_{1m} \sin \theta$$

and the expression for i_2 is

$$i_2 = I_{2m} \sin (\theta - \phi) \qquad (5.16)$$

The negative sign associated with ϕ in equation 5.16 implies that i_2 lags behind i_1 .

5.8 Phasor Diagrams

As has been shown above, sinusoidal alternating quantities can be represented by rotating lines whose length is equal to the maximum value of the quantity concerned. Since we are more frequently concerned with r.m.s. values, it is convenient to represent the r.m.s. values of the quantities together with their phase relationships on diagrams similar to figure 5.6. Such diagrams are known as *phasor diagrams,* and lines drawn on them are known as *phasors.*

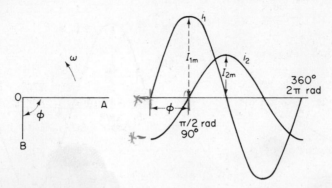

Figure 5.6   Representation of the phase difference between two sinusoidal waveforms

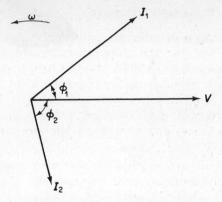

Figure 5.7 A phasor diagram

In figure 5.7 the voltage V is represented by a horizontal line; that is, the length of phasor V is equal to the r.m.s. value of the circuit voltage and, since it is on the horizontal axis, the instantaneous value of the voltage waveform is zero when $\theta = 0$ (see also figure 5.5). Phasor I_1 represents the r.m.s. value of current I_1, which leads V by angle ϕ_1. Also, phasor I_2 represents the r.m.s. value of I_2, which lags behind V by angle ϕ_2.

5.9 Addition and Subtraction of Phasors

The *addition of two phasors* **OA** and **OB** is illustrated in figure 5.8. The phasor sum is obtained graphically by completing the parallelogram OACB, the phasor sum being equal to the diagonal **OC** whose phase angle with respect to the horizontal is ϕ. The values of OC and ϕ are calculated from a knowledge of the horizontal and vertical components of **OA** and **OB** as follows. If h_a and h_b are the horizontal components of **OA** and **OB**, and v_a and v_b are the respective vertical components, then the horizontal component h_c of **OC** is

$$h_c = h_a + h_b$$

and the vertical component v_c of **OC** is

$$v_c = v_a + v_b$$

By Pythagoras' theorem, the *magnitude* or *modulus* of line **OC**, written as | OC |, is

$$| OC | = \sqrt{(h_c^2 + v_c^2)} \tag{5.17}$$

and the phase angle ϕ is

$$\phi = \tan^{-1}(v_c/h_c) \tag{5.18}$$

To *subtract* one phasor from another, it is necessary to add the 'negative' equivalent of the phasor to be subtracted. If, in figure 5.9 we subtract phasor **B**

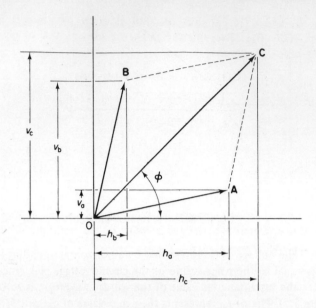

Figure 5.8   Addition of two phasor quantities

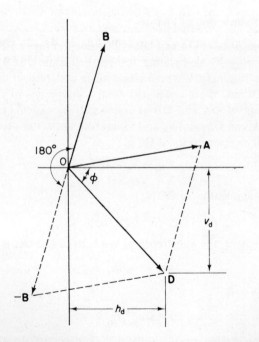

Figure 5.9   Subtraction of two phasor quantities

from phasor **A**, and a phase shift of $180°$ is given to **OB** this produces the equivalent of −**OB**. The latter phasor is then added to **OA** to give the *phasor difference* **OD** where

$$OD = OA - OB = OA + (-OB)$$

The magnitude and phase angle of **OD** are calculated as follows

$$| OD | = \sqrt{(h_d^2 + v_d^2)}$$
$$\varphi = \tan^{-1} (v_d/h_d)$$

Example 5.5

The following e.m.f.s. are induced in coils which are connected in series so that the e.m.f.s are additive. Determine the r.m.s. value of the resultant voltage wave and also its phase angle with respect to e_1.

$$e_1 = 20 \sin \omega t \qquad e_2 = 24 \sin (\omega t + \pi/6)$$
$$e_3 = 24 \cos \omega t \qquad e_4 = 16 \sin (\omega t - 45°)$$

Solution

The phasor of the resultant e.m.f. is determined graphically in figure 5.10. Readers will note that $\pi/6 = 30°$, and that $\cos \omega t = \sin (\omega t + 90°)$; hence e_2 leads e_1 by $30°$ and e_3 leads e_1 by $90°$. E.M.F. e_4 lags behind e_1 by $45°$. In the figure, e.m.f.s e_1 and e_4 are added together first; the resultant is then added to e_2 by completing

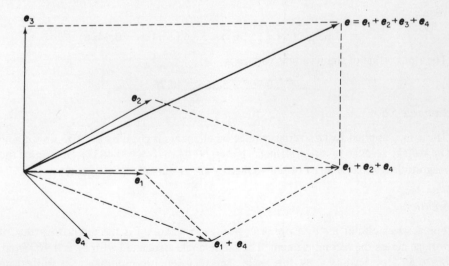

Figure 5.10

another parallelogram to give $e_1 + e_2 + e_4$. This phasor is added to e_3 to give the resultant voltage

$$e = e_1 + e_2 + e_3 + e_4$$

The magnitude and phase angle of e are calculated as follows

horizontal component of $e_1 = 20 \cos 0° = 20$ V

horizontal component of $e_2 = 24 \cos 30° = 24 \times 0.866 = 20.78$ V

horizontal component of $e_3 = 24 \cos 90° = 0$

horizontal component of $e_4 = 16 \cos (-45°) = 16 \times 0.707 = 11.31$ V

total horizontal component $= e_h = 20 + 20.78 + 0 + 11.31 = 52.09$

vertical component of $e_1 = 20 \sin 0° = 0$ V

vertical component of $e_2 = 24 \sin 30° = 24 \times 0.5 = 12$ V

vertical component of $e_3 = 24 \sin 90° = 24$ V

vertical component of $e_4 = 16 \sin (-45°) = 16 \times (-0.707) = -11.31$ V

total vertical component $= e_v = 0 + 12 + 24 - 11.31 = 24.69$

From equation 5.17, the magnitude of e is

$$|e| = \sqrt{(e_h^2 + e_v^2)} = \sqrt{(52.09^2 + 24.69^2)} = 57.65 \text{ V}$$

and from equation 5.18

$$\phi = \tan^{-1} (e_v/e_h) = \tan^{-1} (24.69/52.09)$$
$$= 25.36° \text{ or } 0.4426 \text{ rad} \quad (e \text{ leading } e_1 \text{ by } 25.36°)$$

Hence

$$e = 57.65 \sin (\omega t + 25.36°) = 57.65 \sin (\omega t + 0.4426)$$

The r.m.s. value of the resultant voltage is

$$E = 0.707 \times 57.65 = 40.76$$

Example 5.6

The e.m.f. applied to two series-connected elements is given by $e = 100 \sin \omega t$, and the voltage across one of the elements is $e_1 = 89.44 \sin (\omega t + 0.4637)$. Determine the magnitude and phase angle of the voltage e_2 across the second element.

Solution

For a series circuit $e = e_1 + e_2$, or $e_2 = e - e_1$, where e_2 is the unknown value of voltage across the second element. The phase angle associated with e_1 is 0.4637 rad, or 26.57°, e_1 leading e by this angle. The phasors corresponding to e and e_1 are shown in figure 5.11. To subtract e_1 from e it is necessary to phase shift e_1 through

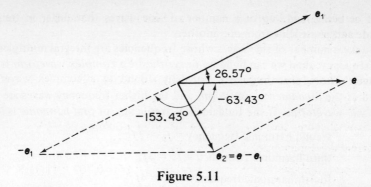

Figure 5.11

$180°$ as shown in the figure. The magnitude and phase angle of e_2 are calculated below

horizontal component of $e = 100 \cos 0° = 100$ V

horizontal component of $-e_1 = 89.44 \cos (26.57° - 180°)$†

$$= 89.44 \cos (-153.43°)$$

$$= - 89.44 \times 0.8944 = - 80 \text{ V}$$

total horizontal component of $e_2 = 100 + (-80) = 20$ V

vertical component of $e = 100 \sin 0° = 0$ V

vertical component of $-e_1 = 89.44 \sin (-153.43°)$

$$= - 89.44 \times 0.4472$$

$$= - 40 \text{ V}$$

total vertical component of $e_2 = 0 + (-40) = - 40$ V

Hence

$$|e_2| = \sqrt{[20^2 + (-40)^2]} = 44.72 \text{ V}$$

and its r.m.s. value is

$$E_2 = 0.707 \times 44.72 = 31.62 \text{ V}$$

Also

$$\phi_2 = \tan^{-1} (-40/20) = \tan^{-1} (-2) = - 63.43°$$

That is, e_2 lags behind e by $63.43°$.

5.10 Harmonics

By far the most important waveform in electrical engineering is the sinusoidal waveform. There are, however, an infinite variety of waveforms which may be

†$26.57° + 180°$ would do equally well.

regarded as being made up of a number of sine waves that differ in frequency, magnitude and phase shift from one another.

If we take a number of sine waves whose frequencies are integral multiples of the lowest frequency, then we can build up or *synthesise* a *complex waveform* from the sine waves. The lowest frequency normally found in a complex waveform is described as the *fundamental waveform*, and the higher frequency waves are known as *harmonic waveforms*. If the fundamental frequency or *first harmonic* is f_1, then

$$\text{second harmonic frequency} = f_2 = 2f_1$$

$$\text{third harmonic frequency} = f_3 = 3f_1$$

$$\text{fourth harmonic frequency} = f_4 = 4f_1$$

etc., and the

$$n\text{th harmonic frequency} = f_n = nf_1$$

Certain types of electrical and electronic equipment operating at power-supply frequencies (50 or 60 Hz) act as harmonic signal generators, and produce harmonics up to and including broadcast frequencies. Such devices include fluorescent lamps, thyristors and Zener diodes.

To illustrate the process of waveform synthesis, let us add a third harmonic frequency to its fundamental frequency. This is illustrated in figure 5.12a for the case where the harmonic is in phase with the fundamental, that is, both waveforms pass through zero simultaneously, after which they increase in the same direction. To synthesise the resultant waveform, we add the values of the respective waveforms together. In figure 5.12a the instantaneous value of the resultant wave at

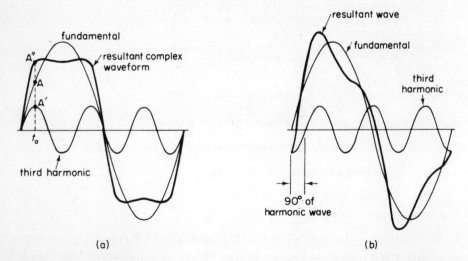

Figure 5.12   Two waveforms that are generated by the addition of a fundamental frequency and its third harmonic

time t_a is A″, and is obtained by adding together the instantaneous values A and A′ of the fundamental and third harmonic waves, respectively. The phase relationship of the harmonics to the fundamental plays an important part in the *shape* of the resultant wave, since, if the third harmonic lags behind the fundamental by 90° (see figure 5.12b), the resultant waveshape differs from that with any other phase relationship. This effect is purely visual, and if both the resultant waveforms in figure 5.12 were reproduced on independent loudspeakers, it would be impossible to detect the effect of the difference in phase shift.

In some instances we may know the shape of the resultant wave but we may want to know which frequencies are present, together with their magnitudes and phase shifts. The procedure of breaking down a complex wave into its constituent components is known as *waveform analysis*. From the foregoing, complex periodic alternating waveforms consist of a number of sine waves, and a mathematical expression giving the instantaneous value of the resultant voltage is

$$v = V_{1m} \sin(\omega t + \phi_1) + V_{2m} \sin(2\omega t + \phi_2) + \ldots + V_{nm} \sin(n\omega t + \phi_n) + \ldots$$

where V_{nm} is the maximum value of the nth harmonic voltage, ϕ_n is the phase relationship of the nth harmonic with respect to a reference signal, and $\omega = 2\pi f$ where f is the fundamental frequency.

Summary of essential formulae

Frequency: $f = 1/T$ hertz

Instantaneous value of e.m.f.: $e = E_m \sin\theta$ volts

Instantaneous value of current: $i = I_m \sin\theta$ amperes

For n equidistant mid-ordinates taken over a half-cycle:
average value $= I_{av} = (i_1 + i_2 + \ldots + i_n)/n$ amperes

R.M.S. value: $I = \sqrt{[i_1^2 + i_2^2 + \ldots + i_n^2)/n]}$ amperes

Form factor = r.m.s. value/average value = I/I_{av}

Peak factor or crest factor = maximum value/r.m.s. value = I_m/I

Sinusoidal waves: average value = $I_{av} = 0.637 I_m$

r.m.s. value = $I = 0.707 I_m$

form factor = 1.11

peak factor = 1.414

Complex waves: $v = V_{1m} \sin(\omega t + \phi) + V_{2m} \sin(2\omega t + \phi_2) + \ldots$
$+ V_{nm} \sin(n\omega t + \phi_n) + \ldots$

PROBLEMS

5.1 The equation relating a sinusoidal alternating voltage (V) with time (s) is $v = 28.28 \sin 628.3t$ V. Determine (a) the r.m.s. value of the voltage, (b) the mean value of the voltage and (c) the frequency and periodic time of the wave.
[(a) 20 V; (b) 18 V; (c) 100 Hz, 0.01 s]

5.2 A circuit is supplied with a sinusoidal current at a frequency of 50 Hz, the r.m.s. value of the current being 5 A. Calculate (a) the peak value of the current and (b) the length of time after passing through zero current it takes for the instantaneous value of current to be 4 A (i) for the first time, (ii) for the second time.
[(a) 7.07 A; (b) (i) 1.91 ms, (ii) 18.09 ms]

5.3 The instantaneous values of an alternating current waveform taken over one half-cycle are as follows.

Time (ms)	0	1	2	3	4	5	6	7	8
Current (A)	0	92	160	200	86	132	100	64	0

Determine (a) the frequency of the waveform, (b) its mean value, (c) its r.m.s. value, (d) its form factor and (e) its crest factor.
[(a) 62.5 Hz; (b) 104.25 A; (c) 114 A; (d) 1.09; (e) 1.16]

5.4 Determine the magnitude and phase angle of the resultant of the following e.m.f.s.

$$e_1 = 10 \sin \omega t \qquad\qquad e_2 = 15 \cos \omega t$$
$$e_3 = 25 \sin (\omega t - \pi/3) \qquad e_4 = 15 \cos (\omega t + 25°)$$

[peak value = 17.59 V; phase angle = 23.27° (0.406 rad)]

5.5 Two single-phase alternators, both generating a supply at the same frequency, are connected in series with one another. If the r.m.s. voltages are 300 and 400 V, respectively, and the phase angle difference between them is 20°, determine the resultant r.m.s. voltage and its phase angle with respect to the component voltages.
[687.6 V; 10.63°]

6 Single-phase Alternating Current Circuits

By definition, alternating quantities vary continuously not only in magnitude but also in direction. This raises some problems when attempting to draw 'direction' arrows on circuit diagrams for such quantities as voltage and current. The notation used in a.c. circuits follows the general principles laid down in chapter 1 for d.c. circuits as follows.

Current – The direction of current flow is indicated by an arrow pointing in the *assumed instantaneous direction* of current flow, the arrow being *on* the branch.
e.m.f. and p.d. – The polarity of the e.m.f. or p.d. between two points is indicated by an arrow between the points which is off the circuit. The arrowhead of the potential arrow points towards the point which, *instantaneously*, is *assumed* to have the more positive potential.

6.1 Circuit Containing Resistance Only

In the circuit in figure 6.1a, the instantaneous values of current and voltage are related by the equation

$$i = \frac{v}{R} \text{ amperes} = \frac{V_m}{R} \sin \omega t = I_m \sin \omega t \tag{6.1}$$

The equation applies *at all instants of time*, and the current and voltage waveforms are *in phase* with one another, as illustrated in figure 6.1b. Also, from equation 6.1 the relationship between the two maximum values is

$$I_m = V_m/R \quad \text{amperes} \tag{6.2}$$

and the r.m.s. current I is

$$I = V/R \quad \text{amperes} \tag{6.3}$$

where V is the r.m.s. value of the voltage across the resistor.

Since the two waveforms are in phase with one another, the phasor diagram is represented by two lines lying on one another (see figure 6.1c).

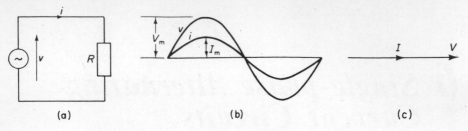

Figure 6.1   An a.c. circuit containing resistance only

Example 6.1

Calculate the r.m.s. value of current in an a.c. circuit whose resistance is 10 Ω, the supply voltage being $v = 28.28 \sin \omega t$.

Solution

$$V = V_m/\sqrt{2} = 28.28/1.414 = 20 \text{ V}$$
$$I = V/R = 20/10 = 2 \text{ A}$$

6.2 Circuit Containing Inductance Only

Consider the circuit in figure 6.2a in which a resistanceless inductor is connected to an alternating supply. From the work in chapter 3, a 'back' e.m.f. is induced in the coil whenever the current flowing through the coil changes value. That is, the net e.m.f. acting in the circuit is

$$v - \text{back e.m.f.} = v - L\frac{di}{dt}$$

Hence the instantaneous current in the circuit is

$$i = \left(v - L\frac{di}{dt}\right)/R$$

or

$$iR = v - L\frac{di}{dt}$$

But, since $R = 0$

$$v = L\frac{di}{dt} \tag{6.4}$$

Therefore the waveforms of v and $L\,di/dt$ are in phase with one another and are equal in magnitude, shown in the upper waveform diagram in figure 6.2b. Since L is

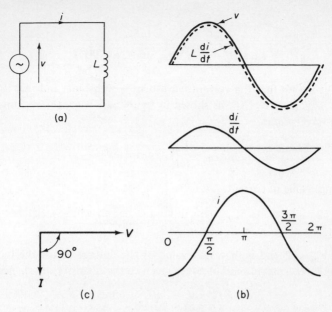

(a)

(c) (b)

Figure 6.2 An a.c. circuit containing a pure inductance

simply a number, then the wave form of di/dt is also in phase with that of v. The shape and phase relationships of the current waveform with respect to the voltage wave are deduced by integrating the di/dt waveform with respect to time. That is

$$i = \int (di/dt) \, dt$$

Since the di/dt waveform is sinusoidal, then the current waveform is a ($-$ cosine) wave, that is, *the current lags behind the voltage by 90°*. Consequently the phasor diagram for the circuit is as shown in figure 6.2c.

The magnitude of the current can be determined from equation 6.4 as follows

$$v = L \, di/dt$$

or

$$di = \frac{1}{L} v \, dt = \frac{1}{L} V_m \sin \omega t \, dt$$

hence

$$i = \frac{V_m}{L} \int \sin \omega t \, dt = -\frac{V_m}{\omega L} \cos \omega t$$

but

$$-\cos \omega t = \sin (\omega t - 90°)$$

therefore

$$i = \frac{V_m}{\omega L} \sin(\omega t - 90°) = I_m \sin(\omega t - 90°) \qquad (6.5)$$

Equation 6.5 shows that the current waveform is sinusoidal and that it lags behind the voltage waveform by 90°, as shown in figure 6.3. Also, the maximum value of the current waveform is

$$I_m = \frac{V_m}{\omega L} \quad \text{amperes} \qquad (6.6)$$

and the r.m.s. value of current, I, is

$$I = \frac{V}{\omega L} = \frac{V}{2\pi f L} \quad \text{amperes} \qquad (6.7)$$

where $V = V_m/\sqrt{2}$, and is the r.m.s. value of the voltage across the inductor. The name *inductive reactance*, symbol X_L, is given to the quantity ωL, hence

$$X_L = \omega L = 2\pi f L \quad \text{ohms} \qquad (6.8)$$

and in a purely inductive circuit it is this quantity which limits the magnitude of the current.

Equation 6.8 shows that the circuit reactance is proportional to frequency, having zero value at zero frequency and having a high value at a high frequency. A graph showing the effect of frequency on both the inductive reactance and on the circuit current is shown in figure 6.4.

Readers approaching the subject of a.c. circuits for the first time may be puzzled by the fact that the current lags behind the voltage across the inductor by a quarter of a cycle and that, in an apparently resistanceless circuit, the magnitude of the

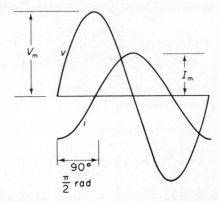

Figure 6.3   In a circuit containing a pure inductance only, the current lags behind the voltage by 90°

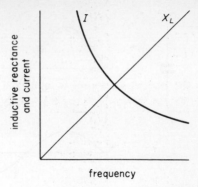

inductive reactance and current

frequency

Figure 6.4 Variation of reactance and current with frequency in a circuit containing a pure inductance

current is limited to a finite value. The reason for both of these lies in the phenomenon of the induced 'back' e.m.f., as follows. When the supply voltage begins to increase, the current cannot increase in sympathy with it, since a back e.m.f. is induced in the winding of the inductance, and this e.m.f. opposes the change of current. The net effect is that the rise of current lags behind the rise of voltage. Also, since the magnitude of the back e.m.f. is related to the rate of change of current and must be equal in value to the supply voltage, then the magnitude of the current is limited by the above phenomenon.

Example 6.2

A pure inductor of 10 mH inductance is connected to an audio-frequency supply of 25 V r.m.s. at a frequency of 5 kHz. Calculate the reactance of the inductor and the value of the r.m.s. current in the circuit.

Solution

$$X_L = 2\pi fL = 2\pi \times 5 \times 10^3 \times 10 \times 10^{-3} = 314.2 \ \Omega$$
$$I = V/X_L = 25/314.2 = 0.0796 \ \text{A}$$

6.3 Circuit Containing Pure Capacitance Only

It was shown in chapter 4 that the equation for the current in a capacitor is

$$i = C\frac{\mathrm{d}v}{\mathrm{d}t} \tag{6.9}$$

where $\mathrm{d}v/\mathrm{d}t$ is the rate of change of voltage across the capacitor. For the circuit in figure 6.5a, the waveforms of the current i and of $C\,\mathrm{d}v/\mathrm{d}t$ are superimposed upon one another. Since C is simply a number, the waveform of $\mathrm{d}v/\mathrm{d}t$ is in phase with the

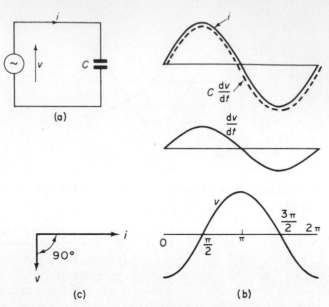

Figure 6.5   An a.c. circuit containing a pure capacitance

current wave, shown in the centre waveform in figure 6.5b. The shape of the waveform of the voltage across the capacitor is obtained by integrating the waveshape of the dv/dt curve; this gives the lower curve in figure 6.5b. Comparing the current and voltage waveforms, we see that *the circuit current leads the voltage across the capacitor by 90°*. The phasor diagram corresponding to the waveform diagrams for i and v is shown in figure 6.5c.

The scientific relationship between the current and voltage for a sinusoidal supply is derived from equation 6.9 as follows

$$i = C\frac{dv}{dt} = C\frac{d}{dt}(V_m \sin \omega t) = \omega C V_m \cos \omega t$$

$$= \omega C V_m \sin(\omega t + 90°) = I_m \sin(\omega t + 90°) \tag{6.10}$$

Equation 6.10 indicates that the current has a maximum value of $\omega C V_m$, and that the current leads the voltage across the capacitor by 90°. The maximum value of current is therefore

$$I_m = \omega C V_m = V_m/X_C \quad \text{amperes}$$

where X_C is the *capacitive reactance* of the capacitor and has the dimensions of ohms. That is

$$X_C = 1/\omega C = 1/2\pi f C \quad \text{ohms} \tag{6.11}$$

The phase relationship between the voltage across the capacitor and the current through it is illustrated in figure 6.6.

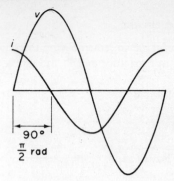

Figure 6.6 In a circuit containing a pure capacitance the current leads the voltage by 90°

From equation 6.11 the capacitive reactance varies inversely with frequency, and its value is infinity at zero frequency and zero at infinite frequency. The resulting graphs of capacitive reactance and circuit current plotted to a base of frequency in a circuit containing only pure capacitance are shown in figure 6.7.

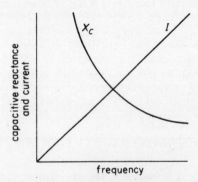

Figure 6.7 Variation of reactance and current with frequency in a circuit containing a pure capacitance

Example 6.3

A 10 μF capacitor is used in an electrical circuit which operates at a frequency of 1.5 kHz. Calculate the reactance of the capacitor at this frequency. If the voltage across the capacitor is 15 V r.m.s., determine the value of the circuit current.

Solution

$$X_C = 1/2\pi f C = 1/2\pi \times 1.5 \times 10^3 \times 10 \times 10^{-6}$$
$$= 10.61 \ \Omega$$
$$I = V/X_C = 15/10.61 = 1.41 \ A$$

A useful mnemonic in a.c. circuits

The word **CIVIL** is a useful aid in remembering the phase relationship between the voltage and the current in capacitive and in inductive circuit elements as follows:

$$
\begin{array}{c}
\underbrace{\text{In C, I leads V}} \\
\overbrace{\text{CIVIL}} \\
\underbrace{\text{V leads I in L}}
\end{array}
$$

6.4 Series Circuit Containing Resistance and Inductance

In the RL series circuit, figure 6.8a, the same current flows through R and L, and the applied voltage is equal to the *phasor sum* of the p.d.s across R and L. The voltage across R is

$$V_R = IR$$

and is in phase with the current (see figure 6.8b). The voltage across L is

$$V_L = IX_L$$

which, from the work in section 6.2, leads the current through it by 90°. The supply voltage V is equal to the phasor sum of V_R and V_L, as shown in figure 6.8b, and its magnitude is

$$V = \sqrt{[(IR)^2 + (IX_L)^2]} = I\sqrt{(R^2 + X_L{}^2)}$$

The quantity $\sqrt{(R^2 + X_L{}^2)}$ is known as the *impedance,* symbol Z, of the RL series circuit, whence

$$V = IZ \tag{6.12}$$

The phase angle between the supply voltage and the circuit current is determined from the *voltage triangle*, figure 6.8c, which is derived from the information in figure 6.8b.

$$\phi = \tan^{-1}\frac{IX_L}{IR} = \tan^{-1}\frac{X_L}{R} = \tan^{-1}\frac{\omega L}{R} \tag{6.13}$$

If the length of each side of the voltage triangle is divided by I, the *impedance triangle* (figure 6.8d) remains. The impedance triangle is more useful in some respects than the voltage triangle since it is geometrically similar to the voltage

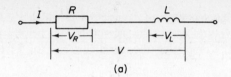

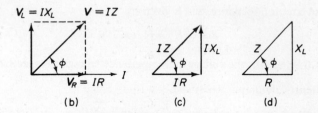

(b)   (c)   (d)

Figure 6.8   *RL* series circuit

triangle, but is independent of the applied voltage. The values of the impedance and phase angle can also be determined from the impedance triangle.

$$\cos \phi = V_R/V = R/Z \qquad (6.14)$$

A practice commonly adopted when drawing phasor diagrams of electrical circuits is to draw the phasor of the quantity that is common to all (or most) components on the horizontal or reference axis. In the case of series circuits this quantity is the current; in parallel circuits it is the voltage.

Example 6.4

In a series *RL* circuit, $R = 15 \ \Omega$ and $L = 0.05$ H. If the circuit is connected to a 100-V r.m.s., 50-Hz supply, calculate the value of the current and of its phase angle with respect to the applied voltage.

Solution

$$X_L = 2\pi f L = 2\pi \times 50 \times 0.05 = 15.7 \ \Omega$$
$$Z = \sqrt{(R^2 + X_L^2)} = \sqrt{(15^2 + 15.7^2)} = 21.7 \ \Omega$$
$$I = V/Z = 100/21.7 = 4.61 \ A$$
$$\phi = \tan^{-1} (X_L/R) = \tan^{-1} (15.7/15) = 46.31° \ (I \text{ lagging behind } V)$$

6.5 Series Circuit Containing Resistance and Capacitance

In the circuit in figure 6.9a, the voltages across the circuit elements are

$$V_R = IR$$
$$V_C = IX_C = I/\omega C$$

144   *Electrical Circuits and Systems*

and are shown on the phasor diagram figure 6.9b, in which V_R is in phase with the current and V_C, the voltage across the capacitor, lags behind the current by 90° (see also section 6.3). Completing the parellelogram for V, the supply voltage, yields

$$V = \sqrt{[(IR)^2 + (IX_C)^2]} = I\sqrt{(R^2 + X_C^2)} = IZ \tag{6.15}$$

where Z is the circuit impedance, and is given by

$$Z = \sqrt{(R^2 + (1/\omega C)^2)} \tag{6.16}$$

From figures 6.9c and d, the voltage and impedance triangles respectively, it can be seen that

$$\phi = \tan^{-1}\frac{V_C}{V_R} = \tan^{-1}\frac{X_C}{R} \tag{6.17}$$

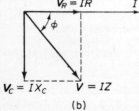

(a)

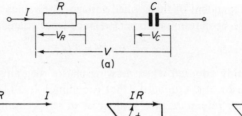

(b)   (c)   (d)

Figure 6.9   *RC* series circuit

and

$$\cos\phi = \frac{V_R}{V} = \frac{R}{Z} \tag{6.18}$$

and that the circuit currents *leads* the applied voltage by ϕ.

Example 6.5

In a series *RC* circuit the voltage across the resistance is found to be 34.2 V when the current flowing is 1.71 A. If the circuit is supplied by a 50 V, 50 Hz source, determine the values of the circuit impedance, the capacitive reactance and the capacitance of the capacitor. Calculate also the phase angle of the circuit.

Solution

$$R = V_R/I = 34.2/1.71 = 20 \ \Omega$$

Since $I = 1.71$ A, then

$$Z = V/I = 50/1.71 = 29.24 \ \Omega$$
$$= \sqrt{(R^2 + X_C^2)}$$

hence

$$X_C = \sqrt{(Z^2 - R^2)} = \sqrt{(29.24^2 - 20^2)} = 21.33 \ \Omega$$
$$= 1/2\pi fC$$

therefore

$$C = 1/2\pi fX_C = 1/(2\pi \times 50 \times 21.33) \ \text{F}$$
$$= 149.2 \ \mu\text{F}$$

and

$$\cos \phi = V_R/V = 34.2/50 = 0.684$$

and

$$\phi = 46.84° \ (I \ \text{leading} \ V)$$

6.6 Series Circuit Containing Resistance, Inductance and Capacitance

The *RLC* series circuit contains all three types of element, and we need to consider three operating conditions, namely when

(a) $X_L > X_C$
(b) $X_L < X_C$
(c) $X_L = X_C$ Resonance

Condition c is a special case known as *resonance* and is dealt with in detail in section 6.7.

(a) $X_L > X_C$

A phasor diagram for the circuit in this condition is shown in figure 6.10b, together with its impedance triangle in figure 6.10c. Since $X_L > X_C$, then $IX_L > IX_C$ or $V_L > V_C$. That is, the circuit appears as an inductive load to the supply, and the current lags behind the voltage by angle ϕ. Inspecting the phasor diagram yields

$$V = \sqrt{[(IR)^2 + (IX_L - IX_C)^2]} = I\sqrt{[R^2 + (X_L - X_C)^2]}$$

$$= I\sqrt{[R^2 + (\omega L - 1/\omega C)^2]} \qquad (6.19)$$

$$= IZ \qquad (6.20)$$

and the circuit impedance is

$$Z = \sqrt{[R^2 + (X_L - X_C)^2]} = \sqrt{[R^2 + (\omega L - 1/\omega C)^2]} \qquad (6.21)$$

Also

$$\phi = \tan^{-1}\ [(V_L - V_C)/V_R] = \tan^{-1}\ [(X_L - X_C)/R] \qquad (6.22)$$

and

$$\cos\phi = \frac{V_R}{V} = \frac{R}{Z} \qquad (6.23)$$

(b) $X_L < X_C$

The phasor diagram for the condition $X_L < X_C$ is shown in figure 6.10d, together with its impedance triangle in figure 6.10e. In this case $V_C > V_L$ and $X_C > X_L$, so

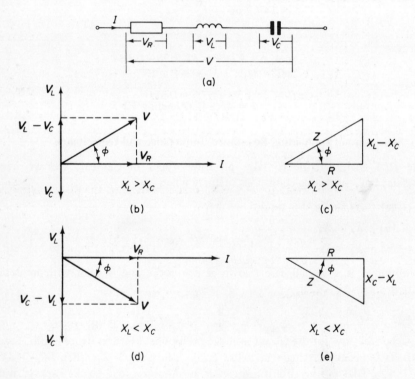

Figure 6.10   *RLC* series circuit

that the current leads the applied voltage. Equations 6.19 to 6.23 above apply in this case with the exception that the terms $(\omega L - 1/\omega C)$ and $(V_L - V_C)$ are replaced by $(1/\omega C - \omega L)$ and $(V_C - V_L)$, respectively. The *general* equations for the *RLC* series circuit are

$$V = \sqrt{[(IR)^2 + (IX_L \sim IX_C)^2]} = IZ \qquad (6.24)$$

$$Z = \sqrt{[R^2 + (X_L \sim X_C)^2]} \qquad (6.25)$$

$$\phi = \tan^{-1}[(X_L \sim X_C)/R] \qquad (6.26)$$

$$\cos \phi = R/Z \qquad (6.27)$$

where the symbol $\sim$ means 'the difference between',

Example 6.6

A current of 5 A at a frequency of 50 Hz flows in a series circuit similar to that in figure 6.10a, which contains a resistance of 11 Ω, an inductance of 0.07 H and a capacitance of 290 μF. Calculate the voltage applied to the circuit and the phase angle between the voltage and the current. Determine also the values of the voltages across the resistor, the inductor and the capacitor.

Solution

$$X_C = 1/2\pi fC = 1/2\pi \times 50 \times 290 \times 10^{-6} = 10.97 \ \Omega$$

$$X_L = 2\pi fL = 2\pi \times 50 \times 0.07 = 22 \ \Omega$$

$$X_L - X_C = 22 - 10.97 = 11.03 \ \Omega$$

$$Z = \sqrt{[R^2 + (X_L \sim X_C)^2]} = \sqrt{(11^2 + 11.03^2)}$$

$$= 15.58 \ \Omega$$

Since $X_L > X_C$, the circuit appears as an inductive load to the power supply and the *current lags behind the supply voltage.*

$$\phi = \tan^{-1}[(X_L \sim X_C)/R] = \tan^{-1}(11.03/11)$$

$$= 45.1° \ (I \text{ lagging } V)$$

The voltages across the various parts of the circuit are

$$V = IZ = 5 \times 15.58 = 77.9 \text{ V}$$

$$V_R = IR = 5 \times 11 = 55 \text{ V}$$

$$V_L = IX_L = 5 \times 22 = 110 \text{ V}$$

$$V_C = IX_C = 5 \times 10.97 = 54.85 \text{ V}$$

The phasor diagram for the circuit is shown in figure 6.11.

Note This circuit raises an interesting point in so far as the value of the voltage across the inductance is greater than the value of the supply voltage! This condition is discussed more fully in section 6.7.

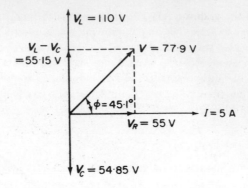

Figure 6.11   Phasor diagram for example 6.6

6.7 Series Resonance

During earlier discussions it was shown that while inductive reactance increased with increasing signal frequency, capacitive reactance reduced with increasing signal frequency. If an inductor and a capacitor are connected in series to a variable frequency supply, there is a frequency known as the *resonant frequency*, f_0 or ω_0, at which the inductive and capacitive reactances are equal to one another. At this frequency the net reactance in the series circuit is zero, when

$$|X_L| = |X_C| \tag{6.28}$$

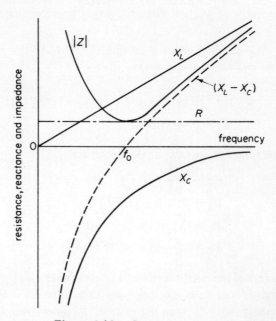

Figure 6.12   Series resonance

Figure 6.12 shows the graphs of X_L, X_C and $(X_L - X_C)$, and at frequency f_0 the value of $(X_L - X_C)$ is zero. The curve showing the modulus of the series circuit impedance, $|Z|$, is also shown. The latter curve has a high value at low frequencies due to the high reactance of the capacitance, and also has a high value at high frequency due to the high reactance of the inductance at these frequencies. The circuit impedance has its minimum value at frequency f_0 when the net reactance is zero and

$$Z = \sqrt{[R^2 + (X_L \sim X_C)^2]} = R \tag{6.29}$$

thus

$$I = V/R \tag{6.30}$$

and

$$\phi = 0° \tag{6.31}$$

that is, the current and voltage are in phase with one another.

The phasor diagram of a series circuit at resonance is shown in figure 6.13. From equation 6.30, the supply voltage is equal to the voltage across the resistor; if the circuit resistance has a low value, the current at resonance is very high. The consequence of the high value of circuit current is that the voltages appearing across both the inductor and the capacitor may be very large, possibly many times the supply voltage. Care has to be taken in series resonant circuits to ensure that the insulation of the inductor and the dielectric strength of the capacitor are adequate to deal with the voltages involved.

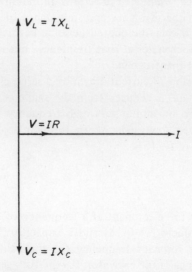

Figure 6.13   Phasor diagram of an *RLC* series circuit at resonance

From the foregoing, when resonance occurs

$$X_L = X_C \quad \text{or} \quad \omega_0 L = 1/\omega_0 C$$

where ω_0 is the resonant frequency in rad/s. Hence

$$\omega_0 = 1/\sqrt{(LC)} \quad \text{rad/s} \tag{6.32}$$

and

$$f_0 = 1/2\pi\sqrt{(LC)} \quad \text{Hz} \tag{6.33}$$

A factor known as the *Q-factor* or 'quality factor' indicates the voltage magnification across either the inductor or the capacitor at resonance. Hence

$$Q\text{-factor} = \frac{\text{voltage across } L \text{ (or } C\text{) at resonance}}{\text{voltage across } R \text{ at resonance}}$$

$$= \frac{I\omega_0 L}{IR} = \frac{\omega_0 L}{R} = \frac{2\pi f_0 L}{R} \tag{6.34}$$

Since $\omega_0 = 1/\sqrt{(LC)}$, then

$$Q\text{-factor} = \frac{1}{R}\sqrt{\left(\frac{L}{C}\right)} \tag{6.35}$$

In some cases the so-called *Q-factor of the coil* is quoted in technical literature. This arises from the fact that the resistance and inductance of the coil are inseparable, and its Q-factor ($= \omega L/R$) may be quoted at some particular frequency ω. This is not to say that the coil resonates at that frequency, but merely gives the ratio of reactance to resistance at that frequency.

The variation of the value of current drawn by a series circuit as frequency varies is shown in figure 6.14, and is deduced from the shape of the impedance curve in figure 6.12. Both at very low and at very high frequency the circuit impedance is high, and the current has a low value. At the resonant frequency of the circuit the current has a peak value of $I = V/R$.

Example 6.7

A series circuit is found to be resonant at a frequency of 50 Hz, and consists of a resistor of 20 Ω, an inductor of 0.3 H and a capacitor. If the supply voltage is 230 V, calculate at the resonant frequency (a) the value of the current in the circuit, (b) the capacitance of the capacitor, (c) the voltages across each of the reactive elements and (d) the Q-factor of the circuit.

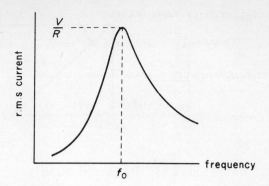

Figure 6.14 Variation of current with frequency in an *RLC* series circuit

Solution

(a) $I = V/R = 230/20 = 11.5$ A
(b) At resonance

$$X_L = 2\pi f_0 L = 2\pi \times 50 \times 0.3 = 94.25 \ \Omega$$

and, at resonance $X_L = X_C$, hence

$$C = 1/2\pi f_0 X_C = 1/2\pi f_0 X_L = 1/2\pi \times 50 \times 94.25 \text{ F} = 33.8 \ \mu\text{F}$$

(c) $V_C = V_L = IX_L = 11.5 \times 94.25 = 1084$ V
(d) Q-factor $= 2\pi f_0 L/R = 94.25/20 = 4.713$

Note The voltage across the reactive elements is 4.713 times greater than the supply voltage!

6.8 Parallel Circuit Containing Resistance and Inductance

In a parallel *RL* circuit — see figure 6.15a — the voltage across both elements is equal to the supply voltage *V*. Consequently, voltage *V* is used as the reference in the phasor diagram, figure 6.15b. The current I_R flowing through the resistive arm, by reason of the work in section 6.1, is in phase with the supply voltage. Also, the current I_L flowing in the inductive branch lags behind the applied voltage by 90° (see section 6.2). The current *I* drawn from the supply is the *phasor sum* of I_R and I_L, as shown in figure 6.15b; *the resultant current lags behind the voltage applied to the parallel RL circuit.* Now

$$I_R = V/R \qquad\qquad (6.36)$$

and

$$I_L = V/X_L = V/\omega L = V/2\pi f L \qquad\qquad (6.37)$$

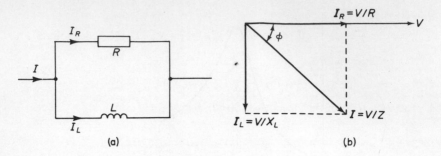

Figure 6.15   *RL* parallel circuit

The magnitude of the current *I* drawn by the circuit is

$$|I| = \sqrt{(I_R^2 + I_L^2)} = V/Z \qquad (6.38)$$

where *Z* is the magnitude of the circuit impedance and

$$\phi = \tan^{-1}(I_L/I_R) \qquad (6.39)$$

also

$$\cos\phi = \frac{I_R}{I} = \frac{V/R}{V/Z} = \frac{Z}{R} \qquad (6.40)$$

Example 6.8

In an *RL* parallel circuit of the type in figure 6.15, $R = 10\ \Omega$ and $L = 0.05$ H. If the circuit is energised by a 220 V r.m.s., 50 Hz supply, determine the value of the current drawn from the supply and its phase angle with respect to the supply voltage. Calculate also the magnitude of the impedance of the circuit.

Solution

$$X_L = 2\pi fL = 2\pi \times 50 \times 0.05 = 15.71\ \Omega$$

therefore

$$I_L = V/X_L = 220/15.71 = 14\ A$$

and

$$I_R = V/R = 220/10 = 22\ A$$

The magnitude of the current drawn from the supply is

$$|I| = \sqrt{(I_R^2 + I_L^2)} = \sqrt{(22^2 + 14^2)} = 26.08\ A$$

From equation 6.39

$$\phi = \tan^{-1}(I_L/I_R) = \tan^{-1}(14/22) = 32.47° \ (I \text{ lagging behind } V)$$

and from equation 6.38

$$Z = V/I = 220/26.08 = 8.44 \ \Omega$$

The phasor diagram of the circuit is shown in figure 6.16.

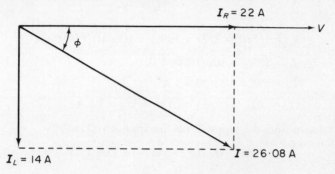

Figure 6.16

6.9 Parallel Circuit Containing Resistance and Capacitance

A circuit diagram of a parallel RC circuit together with its phasor diagram is shown in figure 6.17. It differs from the parallel RL circuit only in that the current in the reactive branch leads the supply voltage by 90°. The relevant equations for the circuit are

$$I_R = V/R \tag{6.41}$$

$$I_C = V/X_C = V/(1/\omega C) = \omega CV = 2\pi f CV \tag{6.42}$$

$$|I| = \sqrt{(I_R^2 + I_C^2)} = V/Z \tag{6.43}$$

$$\phi = \tan^{-1}(I_C/I_R) = \tan^{-1}(R/X_C) = \tan^{-1}\omega CR \tag{6.44}$$

$$\cos\phi = I_R/I = Z/R \tag{6.45}$$

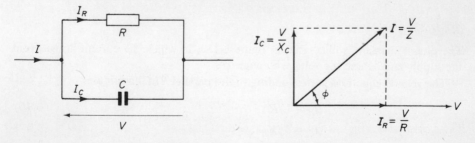

Figure 6.17 *RC* parallel circuit

Example 6.9

In a circuit of the type in figure 6.17, $R = 20\ \Omega$ and $C = 10\ \mu\text{F}$. If the circuit is energised at a voltage of 100 V r.m.s., 159.1 Hz, calculate the reactance of the capacitor, the current flowing in each branch of the circuit and the value of the current drawn from the supply. Determine also the phase angle of the current with respect to the supply voltage.

Solution

$$X_C = 1/2\pi fC = 1/2\pi \times 159.1 \times 10 \times 10^{-6} = 100\ \Omega$$

$$I_C = V/X_C = 100/100 = 1\ \text{A}$$

$$I_R = V/R = 100/20 = 5\ \text{A}$$

$$|I| = \sqrt{(I_R{}^2 + I_C{}^2)} = \sqrt{(5^2 + 1^2)} = 5.1\ \text{A}$$

The current drawn from the supply leads the applied voltage by

$$\phi = \tan^{-1}\ (I_C/I_R) = \tan^{-1}\ (1/5) = 11.3°$$

6.10 Parallel Circuit Containing Resistance, Inductance and Capacitance

For this type of circuit we need to consider the following three conditions

(a) $I_C > I_L$ that is, $X_C < X_L$

(b) $I_C < I_L$ that is, $X_C > X_L$

(c) $I_C = I_L$ that is, $X_C = X_L$

Condition c is the special case of *parallel resonance*, and is dealt with in section 6.11.

(a) $I_C > I_L$

The phasor diagram for this condition is shown in figure 6.18b. In this case the current *I* drawn from the supply leads the applied voltage by angle ϕ.

(b) $I_C < I_L$

The phasor diagram is illustrated in figure 6.18c, in which the current drawn from the supply lags behind the voltage by angle ϕ.

The general equations corresponding to the parallel *RLC* circuit are

$$I_R = V/R \tag{6.46}$$

$$I_L = V/X_L \tag{6.47}$$

$$I_C = V/X_C \tag{6.48}$$

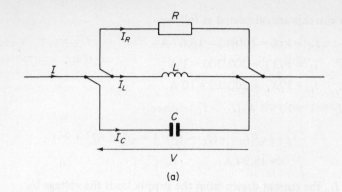

(a)

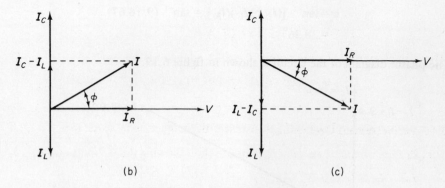

(b)   (c)

Figure 6.18   *RLC* parallel circuit

$$I = \sqrt{[I_R{}^2 + (I_L \sim I_C)^2]} = V/Z \qquad (6.49)$$

$$\phi = \tan^{-1}\ [(I_L \sim I_C)/I_R] \qquad (6.50)$$

$$\cos \phi = I_R/I = Z/R \qquad (6.51)$$

Example 6.10

A circuit of the type in figure 6.18a, in which $R = 12\ \Omega$, $L = 100$ mH and $C = 50\ \mu$F, is supplied by a 200 V r.m.s., 318.3 Hz source. Determine the value of the current in each branch of the circuit, the total current drawn from the supply and its phase angle.

Solution

$$X_L = 2\pi f L = 2\pi \times 318.3 \times 100 \times 10^{-3} = 200\ \Omega$$
$$X_C = 1/2\pi f C = 1/2\pi \times 318.3 \times 50 \times 10^{-6} = 10\ \Omega$$

The branch currents are calculated as follows

$$I_R = V/R = 200/12 = 16.67 \text{ A}$$
$$I_L = V/X_L = 200/200 = 1 \text{ A}$$
$$I_C = V/X_C = 200/10 = 10 \text{ A}$$

Now $I_L \sim I_C = 1 \sim 10 = 9 \text{ A}, (I_C > I_L)$, hence

$$|I| = \sqrt{[I_R^2 + (I_L \sim I_C)^2]} = \sqrt{(16.67^2 + 9^2)}$$
$$= 18.94 \text{ A}$$

Since $I_C > I_L$, the current drawn from the supply leads the voltage by

$$\phi = \tan^{-1} [(I_L \sim I_C)/I_R] = \tan^{-1} (9/16.67)$$
$$= 28.36°$$

The phasor diagram of the circuit is shown in figure 6.19.

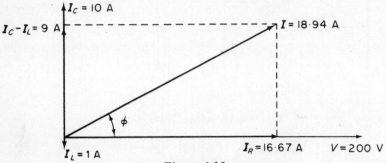

Figure 6.19

6.11 Parallel Resonance

Since the resistance and inductance of a coil cannot be physically separated, the circuit in figure 6.20a is that of a practical parallel circuit. At the resonant frequency, f_0 or ω_0, of the circuit the quadrature components of the currents in the two branches are equal to one another. That is, the quadrature components cancel each other out, and the current I drawn from the supply is in phase with the supply voltage. The phasor diagram of the circuit at resonance is shown in figure 6.20b. The angle ϕ_L in the phasor diagram is the phase angle of the inductive branch, so that

$$\sin \phi_L = \frac{\text{reactance of the coil at resonance}}{\text{impedance of the coil at resonance}} = \frac{X_L}{Z}$$

$$= \frac{\omega_0 L}{\sqrt{[R^2 + (\omega_0 L)^2]}}$$

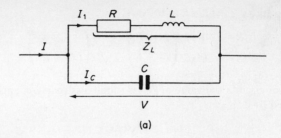

(a)

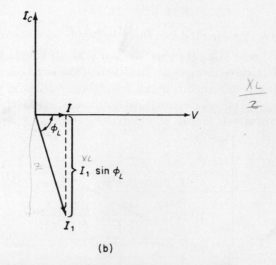

(b)

Figure 6.20 Resonance in a parallel circuit

Now the current in the inductive branch is $I_1 = V/Z_L$, and its quadrature component at resonance is

$$I_1 \sin \phi_L = \frac{V}{Z_L} \times \frac{X_L}{Z_L} = \frac{VX_L}{Z_L{}^2}$$

At resonance $I_C = I_1 \sin \phi_L$ and, since $I_C = V/X_C$ then

$$\frac{V}{X_C} = \frac{VX_L}{Z_L{}^2}$$

or

$$Z_L{}^2 = X_L X_C$$

that is

$$R^2 + (\omega_0 L)^2 = \omega_0 L \times \frac{1}{\omega_0 C} = \frac{L}{C}$$

Solving for ω_0 yields

$$\omega_0 = \frac{1}{L}\sqrt{\left(\frac{L}{C} - R^2\right)} = \sqrt{\left(\frac{1}{LC} - \frac{R^2}{L^2}\right)} \tag{6.52}$$

In many practical circuits the value of R^2 is much less than L/C, and equation 6.52 may be simplified to

$$\omega_0 \approx 1/\sqrt{(LC)} \tag{6.53}$$

or

$$f_0 \approx 1/2\pi\sqrt{(LC)} \tag{6.54}$$

Readers will note that equations 6.53 and 6.54 are identical to equations 6.32 and 6.33 which relate to the resonant condition of the series circuit.

At resonance the current drawn from the supply is in phase with the supply voltage, and the circuit presents a purely resistive load at this frequency. The value of this resistance is known as the *dynamic resistance*, R_D, and is calculated as follows

$$R_D = \frac{V}{I} = \frac{V}{I_1 \sin\phi_L / \tan\phi_L} = \frac{V\tan\phi_L}{I_C} = \frac{V\tan\phi_L}{V/X_C}$$

$$= X_C \tan\phi_L = \frac{1}{\omega_0 C} \times \frac{\omega_0 L}{R} = \frac{L}{CR} \quad \text{ohms} \tag{6.55}$$

but, since $\omega_0^2 = 1/LC$, then

$$R_D = \frac{L}{CR} = \frac{\omega_0^2 L^2}{R} = \frac{1}{\omega_0^2 C^2 R} \tag{6.56}$$

Readers will note from the above equations that when $R = 0$, then $R_D = \infty$! This implies that the smaller the value of R, the smaller the value of current drawn from the supply at resonant frequency. In the limiting case when $R = 0$, no current is drawn from the supply at resonance.

For the above reason, the parallel resonant circuit is known as a *rejector circuit*, since it rejects current at the resonant frequency of the circuit. A graph showing the variation of the current drawn from the supply to a base of frequency is shown in figure 6.21. At zero frequency (d.c.) the effective value of circuit impedance is equal to the resistance of the coil, so that the current has a large value. At frequencies that are well above resonance the reactance of the capacitor is low, once more giving a large value of current. The minimum value of supply current of V/R_D occurs at resonance.

Inspecting the phasor diagram in figure 6.20b reveals that the value of the current I_C which circulates at resonance within the parallel circuit can be many times greater than the value of current I. In fact, the current drawn from the supply simply provides the energy losses within the parallel circuit at resonance. The

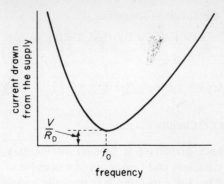

Figure 6.21 Variation of current with frequency in an *RLC* parallel circuit

Q-factor of the parallel circuit is the ratio of the value of the circulating current at resonance to the value of the current drawn from the supply, and is a measure of the *current magnification* within the circuit. Referring to figure 6.20

$$Q = \frac{\text{reactive current}}{\text{supply current}} = \frac{I_C}{I} = \tan \phi = \frac{\omega_0 L}{R} \qquad (6.57)$$

and, since $\omega_0 L = 1/\omega_0 C$, then

$$Q = 1/\omega_0 CR \qquad (6.58)$$

Example 6.11

A coil of resistance 12 Ω and inductance 0.12 H is connected in parallel with a 60 μF capacitor to a 100 V supply. Calculate the value of the resonant frequency of the circuit and determine its dynamic impedance at this frequency. What current is drawn from the supply at resonance? Also compute the Q-factor of the circuit.

Solution

From equation 6.52

$$\omega_0 = \sqrt{\left(\frac{1}{LC} - \frac{R^2}{L^2}\right)} = \sqrt{\left(\frac{1}{0.12 \times 60 \times 10^{-6}} - \frac{12^2}{0.12^2}\right)}$$

$$= \sqrt{(138\,888 - 10\,000)} = 359 \text{ rad/s}$$

$$f_0 = \omega_0/2\pi = 57.1 \text{ Hz}$$

Note In the above calculation, $(R/L^2) \ll (1/LC)$ and $\omega_0 \approx 1/\sqrt{(LC)}$. From equation 6.55

$$R_{\rm D} = L/CR = 0.12/(60 \times 10^{-6} \times 12) = 166.7 \ \Omega$$

The current drawn by the circuit at resonance is

$$I = V/R_{\mathrm{D}} = 100/166.7 = 0.6 \text{ A}$$

and

$$Q\text{-factor} = \omega_0 L/R = 359 \times 0.12/12 = 3.59$$

6.12 Series–Parallel *RLC* Circuits

In practical circuits each branch of a parallel circuit may itself be a complex network of resistors, inductors and capacitors. An example of this kind of circuit is illustrated in figure 6.22a. Since the circuit is basically a parallel circuit, the supply voltage is used as the reference phasor in the phasor diagram, figure 6.22b. The resultant current drawn from the supply is the phasor sum of the currents in the individual branches. A solution for a circuit of this type is illustrated in the following example.

Example 6.12

A circuit of the type in figure 6.22a contains elements with the following values: $R_1 = 10 \ \Omega$, $X_L = 10 \ \Omega$, $R_2 = 20 \ \Omega$, $R_3 = 20 \ \Omega$, $X_C = 20 \ \Omega$. Determine the values of

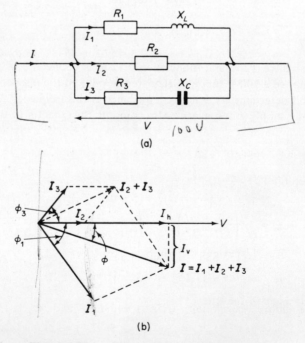

(a)

(b)

Figure 6.22   A complex parallel circuit

each branch current and of the resultant current drawn from the supply.

Solution

To calculate the values and phase angles of the branch currents, it is first necessary to evaluate the magnitudes and phase angles of the branch impedances.

For the upper branch

$$Z_1 = \sqrt{(R_1{}^2 + X_L{}^2)} = \sqrt{(10^2 + 10^2)} = 14.14 \ \Omega$$
$$I_1 = V/Z_1 = 100/14.14 = 7.07 \ \text{A}$$
$$\phi_1 = \tan^{-1} (X_L/R_1) = \tan^{-1} (10/10) = 45° \ (I \text{ lagging } V)$$

For the centre branch

$$Z_2 = R_2 = 20 \ \Omega$$
$$I_2 = V/R_2 = 100/20 = 5 \ \text{A}$$
$$\phi_2 = 0°$$

For the lower branch

$$Z_3 = \sqrt{(R_3{}^2 + X_C{}^2)} = \sqrt{(20^2 + 20^2)} = 28.28 \ \Omega$$
$$I_3 = V/Z_3 = 100/28.28 = 3.535 \ \text{A}$$
$$\phi_3 = \tan^{-1} (X_C/R) = \tan^{-1} (20/20) = 45° \ (I \text{ leading } V)$$

The horizontal component I_h of the resultant current is

$$I_h = I_1 \cos \phi_1 + I_2 \cos \phi_2 + I_3 \cos \phi_3$$
$$= (7.07 \times 0.707) + (5 \times 1) + (3.535 \times 0.707) = 12.5 \ \text{A}$$

and the vertical component I_v of the resultant current is

$$I_v = I_1 \sin \phi_1 + I_2 \sin \phi_2 + I_3 \sin \phi_3$$
$$= [7.07 \times (-0.707)] + (5 \times 0) + (3.535 \times 0.707)$$
$$= - 2.5 \ \text{A}$$

The magnitude of the resultant current is

$$I = \sqrt{[12.5^2 + (-2.5^2)]} = 12.75 \ \text{A}$$

and the phase angle of the current is

$$\phi = \tan^{-1} (I_v/I_h) = \tan^{-1} (-2.5/12.5) = \tan^{-1} - 0.2$$
$$= 11.31° \ (I \text{ lagging } V)$$

6.13 Power Consumed by a Non-reactive Circuit

Since the effective heating value of an alternating current is its r.m.s. value, I, and the effective value of voltage is its r.m.s. value, V, then the power consumed in a resistance R is

$$P = VI = I^2 R \quad \text{watts} \tag{6.59}$$

Expressed in mathematical terms, if the instantaneous values of voltage and current in resistor R are

$$v = V_\mathrm{m} \sin \theta$$

and

$$i = I_\mathrm{m} \sin \theta$$

where $\theta = \omega t$, the instantaneous power consumed is

$$p = vi = V_\mathrm{m} \sin \phi \times I_\mathrm{m} \sin \phi = V_\mathrm{m} I_\mathrm{m} \sin^2 \phi$$

$$= \frac{1}{2} V_\mathrm{m} I_\mathrm{m} (1 - \cos 2\theta) \tag{6.60}$$

The waveform corresponding to the product vi is shown in figure 6.23 and, since v and i are in phase with one another, then the instantaneous power curve always has a positive value except at zero angle and integral multiples of π radians, when it has zero value.

The *average power, P,* consumed by the circuit is determined by evaluating the integral

$$P = \frac{1}{2\pi} \int_0^{2\pi} vi \, \mathrm{d}\theta = \frac{1}{2\pi} \int_0^{2\pi} \frac{1}{2} V_\mathrm{m} I_\mathrm{m} (1 - \cos 2\theta) \, \mathrm{d}\theta$$

$$= \frac{V_\mathrm{m} I_\mathrm{m}}{2} = \frac{\sqrt{2}V \times \sqrt{2}I}{2} = VI \quad \text{watts} \tag{6.61}$$

6.14 Power Consumed by a Circuit Containing Resistance and Reactance

When the current and voltage waveforms are not in phase with one another, which occurs in circuits containing reactance, then the graph of instantaneous power becomes negative during part of the cycle. This is illustrated in figure 6.24 for the case where the current lags behind the voltage. During the interval AB in figure 6.24a, the value of v is positive and that of i is negative, giving the instantaneous product $p = vi$ a negative value. It occurs again during the interval CD when v has a negative value and i has a positive value. The negative part of the power waveform has a physical meaning and is explained below.

During the period BC, power is supplied to the circuit and some of it is stored in the (inductive) reactance. The negative area in period CD represents energy

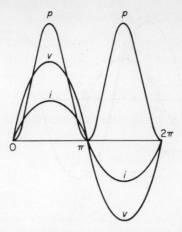

Figure 6.23   Waveforms of voltage, current and power in a resistive circuit

returned from the reactance to the generator. The average power consumed during the whole cycle is proportional to the difference between the positive and negative areas under the instantaneous power curve.

If the angle of lag increases (ϕ increases), then the mean area under the instantaneous power curve is reduced since the 'negative' area increases. Consequently, the average power consumed is reduced as ϕ is increased.

The equations defining the instantaneous values of voltage and current in figure 6.24 are

$$v = V_\mathrm{m} \sin \theta$$

$$i = I_\mathrm{m} \sin (\theta - \phi)$$

and the instantaneous power consumed by the circuit is

$$p = vi = V_\mathrm{m}I_\mathrm{m} \sin \theta \times \sin (\theta - \phi)$$

$$= \frac{1}{2} V_\mathrm{m}I_\mathrm{m} [\cos \phi - \cos (2\theta - \phi)]$$

$$= \frac{1}{2} V_\mathrm{m}I_\mathrm{m} \cos \phi - \frac{1}{2} V_\mathrm{m}I_\mathrm{m} \cos (2\theta - \phi)$$

The average power consumed by the circuit is computed from the integral

$$P = \frac{1}{2\pi} \int_0^{2\pi} p \ \mathrm{d}\theta = \frac{V_\mathrm{m}I_\mathrm{m}}{2} \cos \phi = VI \cos \phi \qquad (6.62)$$

Equation 6.62 can be shown to be valid for both inductive and capacitive circuits.

Example 6.13

Two circuits A and B are energised by a 100 V r.m.s. a.c. supply. The current

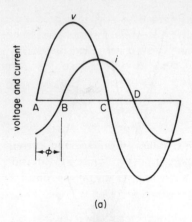

(a)

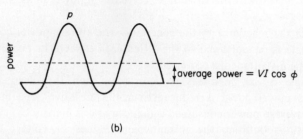

(b)

Figure 6.24   Voltage, current and power waveforms in a circuit containing a
resistor and an inductor

flowing in each circuit is 10 A and the phase angle of circuit A is 0° and that of
circuit B is 60°. Calculate the power consumed by each circuit.

Solution

For circuit A

$$P = VI \cos 0° = 100 \times 10 = 1000 \text{ W}$$

For circuit B

$$P = VI \cos 60° = 100 \times 10 \times 0.5 = 500 \text{ W}$$

6.15 Power Consumed by a Circuit Containing a Pure Reactance Only

From equation 6.62 we see that when $\phi = 90°$ (either lagging or leading), then the
average power consumed by the circuit is

$$P = VI \cos 90° = 0$$

That is, the average power consumed by a pure reactance is zero. The waveform diagrams in figure 6.24 offer an explanation for this phenomenon; when $\phi = 90°$ the average area under the power curve is zero, and the energy consumed by the reactance during each quarter cycle is returned to the generator during the following quarter cycle.

6.16 Power Factor

In a.c. the product of the r.m.s. values of voltage and current is known as the *volt-ampere product* or the *apparent power,* symbol *S*. Multiplying the apparent power by the factor cos ϕ for the circuit gives the power consumed. For this reason, cos ϕ is described as the circuit *power factor.* That is

$$\text{power} = \text{volt amperes} \times \text{power factor}$$

or

$$\text{power factor} = \text{power/volt amperes}$$

that is

$$\cos \phi = P/VI = P/S \tag{6.63}$$

In the case of the series circuit, the power factor is (see figures 6.8, 6.9 and 6.10)

$$\cos \phi = \frac{\text{resistance}}{\text{impedance}} = \frac{R}{Z} \tag{6.64}$$

When the power factor of a series circuit is unity, that is $\phi = 0°$, the power consumed is a maximum. When the phase angle is 90° (either lagging or leading), the power factor is zero and the power consumed is zero.

6.17 Reactive Power

A phasor diagram for an *RL* series circuit is shown in figure 6.25a. The *power triangle* in figure 6.25b is geometrically similar to the voltage triangle and, as we have seen, the apparent power *S* is given by the expression

$$S = VI \quad \text{volt amperes (VA)}$$

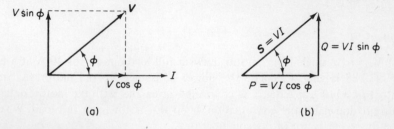

(a)   (b)

Figure 6.25   (a) A phasor diagram and (b) its VA triangle for an a.c. circuit

and the *active power* or *real power*, *P*, is

$$P = VI \cos \phi \quad \text{watts}$$

The product of the current and the reactive component of the voltage ($V \sin \phi$) is known as the *reactive power, Q,* where

$$Q = VI \sin \phi \quad \text{volt amperes reactive (VAr)}$$

6.18 Power Factor Correction

In most forms of electrical power systems the supply voltage remains fairly constant, and the heating effect produced in the equipment is related to the current drawn by the equipment (more correctly, it is related to (current)^2). For this reason many forms of electrical equipment, including transformers and electrical machines, are rated in terms of their volt-ampere capacity. The actual power output (or power consumed) then depends on the power factor at which the machine operates. Thus an alternator rated at 1 MVA can provide, at 1kV, a current of 1000 A whatever the power factor of the connected load. It can, however, only -provide an output of 1 MW at unity power factor. At lower values of power factor the output power is less than 1 MW.

Moreover the heating effect and therefore the frame size of electrical machines is dependent on the value of (current)^2 consumed. Also, for a given amount of power transmitted through the cables and transformers of a transmission system, the lower the value of the power factor the greater the value of current transmitted. Hence, for a given value of transmitted power, a low power factor results in increased power losses.

So important is the value of the power factor of the load that power-supply authorities include a clause in their tariffs which penalises consumers if their power factor falls below a certain level.

The majority of industrial loads have a lagging power factor, and one method of improving the power factor is to connect a capacitor in parallel with the load, as illustrated in the following example. The object of the use of the capacitor is to draw an additional current from the supply which leads the supply voltage by 90°; this current provides some compensation for the lagging current taken by the inductive load.

Example 6.14

A 500 V, 50 Hz single-phase motor draws a full load current of 40 A at a power factor of 0.85 lagging. Calculate the power consumed by the motor. A bank of capacitors of 80 μF capacitance is connected in parallel with the motor; determine the current drawn by the combination when the motor is on full load. What will then be the power factor of the combination?

Solution

The power consumed by the motor is

$$P = VI_1 \cos \phi_1$$

where I_1 is the current drawn by the motor, and ϕ_1 is the phase angle of I_1 with respect to V (see figure 6.26). Hence

$$P = 500 \times 40 \times 0.85 = 17\,000\,\text{W} = 17\,\text{kW}$$

also the component of I_1 which is in phase with the supply voltage is

$$I_1 \cos \phi_1 = 40 \times 0.85 = 34\,\text{A}$$

now

$$\phi_1 = \cos^{-1} 0.85 = 31.79°$$

and the quadrature component or reactive component of I_1 is

$$I_1 \sin \phi_1 = 40 \times \sin 31.79° = 40 \times 0.527 = 21.08\,\text{A}$$

The reactance X_C of the capacitance is

$$X_C = 1/2\pi fC = 1/2(2\pi \times 50 \times 80 \times 10^{-6}) = 39.8\,\Omega$$

and the current the capacitor draws from the supply is

$$I_C = V/X_C = 500/39.8 = 12.56\,\text{A at an angle of } 90° \ (I_C \text{ leading } V)$$

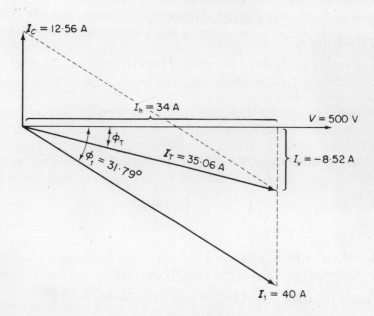

Figure 6.26

The value of the in-phase component I_h of the total current I_T is

$$I_h = I_1 \cos \phi_1 = 34 \text{ A}$$

and the value of the quadrature component I_v is

$$I_v = I_C - I_1 \sin \phi_1 = 12.56 - 21.08 = -8.52 \text{ A}$$

The magnitude of the total current drawn from the supply is

$$|I_T| = \sqrt{(I_h^2 + I_v^2)} = \sqrt{[34^2 + (-8.52)^2]} = 35.05 \text{ A}$$

(*Note* This is a 12.3 per cent reduction of current when compared with the original value of 40 A for the same power consumption.) The power factor of the parallel combination is given by

$$\cos \phi_T = I_T \cos \phi_T / I_T = I_1 \cos \phi_1 / I_T = 34/35.05 = 0.97 \text{ lagging}$$

Summary of essential formulae

Resistive circuit: $V = IR$ and $\phi = 0°$

Purely inductive circuit: $V = IX_L$ and $\phi = 90°$ (current lagging)

$$X_L = \omega L = 2\pi f L$$

Purely capacitive circuit: $V = IX_C$ and $\phi = 90°$ (current leading)

$$X_C = 1/\omega C = 1/2\pi f C$$

R, L and C in series: $V = IZ$

$$Z = \sqrt{[R^2 + (X_L \sim X_C)^2]} = \sqrt{(R^2 + X^2)}$$

$$\phi = \tan^{-1}(X/R) = \cos^{-1}(R/Z)$$

Series resonance: $\omega_0 = 1/\sqrt{(LC)}$ rad/s

$$f_0 = 1/2\pi\sqrt{(LC)}   \text{Hz}$$

$$Q\text{-factor} = \omega_0 L/R = \frac{1}{R}\sqrt{\left(\frac{L}{C}\right)}$$

Parallel resonance: $\omega_0 = \sqrt{\left(\dfrac{1}{LC} - \dfrac{R^2}{L^2}\right)}$ rad/s

$$f_0 = \omega_0/2\pi   \text{Hz}$$

When $(1/LC) \gg (R/L^2)$, then

$$\omega_0 = 1/\sqrt{(LC)}   \text{and}   f_0 = 1/2\pi\sqrt{(LC)}   \text{Hz}$$

dynamic resistance $= R_D = L/CR = \omega_0^2 L^2/R = 1/\omega_0^2 C^2 R$

Q-factor $= \omega_0 L/R = 1/\omega_0 CR$

Apparent power: $S = VI$ volt amperes

Power: $P = I^2 R = VI \cos \phi$ watts

Power factor: $\cos \phi = P/VI = P/S = R/Z$ (in a series circuit)

Reactive power: $Q = VI \sin \phi$ volt amperes reactive (VAr)

PROBLEMS

6.1 Determine the inductive reactance of a coil of inductance 0.05 H when it is connected to a supply frequency of (a) 50 Hz, (b) 60 Hz and (d) 1 MHz.
[(a) 314.2 Ω, (b) 377 Ω; (c) 6.283 MΩ]

6.2 Determine the impedance of the coil in problem 6.1 at each of the frequency values given if its resistance is 200 Ω.
[(a) 372.4 Ω; (b) 426.8 Ω; (c) 6.283 MΩ]

6.3 Coil A of reactance 10.3 Ω and resistance 20 Ω is connected in series with coil B of reactance 15.7 Ω and resistance 10 Ω. Calculate the impedance of the combination.
[39.7 Ω]

6.4 A non-inductive resistor of resistance 15 Ω is connected in series with a coil of resistance 10 Ω and inductance 0.1 H. If the impedance of the circuit is 70 Ω, determine the supply frequency.
[104.06 Hz]

6.5 An inductor of reactance 40 Ω and negligible resistance is connected in series with a resistance of 30 Ω. If the current in the circuit is 10 A, determine (a) the voltage across the inductor, (b) the voltage across the resistor, (c) the supply voltage, (d) the phase angle between the supply voltage and the current, (e) the power factor of the circuit, (f) the power consumed, (g) the VA consumed and (h) the VAr consumed. Draw to scale the phasor diagram of the circuit.
[(a) 400 V, (b) 300 V; (c) 500 V; (d) 53.13° (I lagging V); (e) 0.6; (f) 3 kW; (g) 5 kVA; (h) 4 kVAr]

6.6 A voltage of 100 V d.c. produces a current of 20 A in a coil. When a sinusoidal voltage waveform of 100 V r.m.s. at 50 Hz was applied to the coil the current was 10 A r.m.s. Calculate the resistance, the impedance and the inductance of the coil. Draw the impedance triangle to scale.
[R = 5 Ω; Z = 10 Ω; L = 27.57 mH]

6.7 A series circuit consists of a resistanceless coil of inductance 100 mH and a resistor of resistance 20 Ω. The circuit is connected to the terminals of a single-

phase alternator whose terminal voltage and frequency are proportional to the speed of the shaft. If the alternator output is 100 V at 50 Hz at full speed, calculate (a) the current in the circuit, (b) the phase angle of the circuit and (c) the power consumed when the shaft speed is (i) 120 per cent (ii) 50 per cent of full speed.
[(a) (i) 2.81 A, (ii) 1.97 A; (b) (i) −62.05°, (ii) −38.15°; (c) (i) 157.9 W, (ii) 77.62 W]

6.8 A capacitor of 1.0 μF capacitance is connected to a 10 V r.m.s. a.c. supply of variable frequency. Plot a curve showing how (i) the capacitive reactance, (ii) the current varies with frequency over the range 100 Hz to 1 kHz.

6.9 A series-connected a.c. circuit consists of a capacitor of reactance 10 Ω and a resistor of resistance 15 Ω. Determine (a) the magnitude of the impedance of the circuit. If the supply voltage is 10 V r.m.s, calculate (b) the current in the circuit, (c) the phase angle of the current with respect to the supply voltage, and (d) the power consumed by the circuit. Draw to scale the phasor diagram of the circuit.
[(a) 18.03 Ω; (b) 0.555 A; (c) 33.69° (leading); (d) 4.62 W]

6.10 A non-inductive resistor of resistance 100 Ω is connected in series with a capacitor to a 250-V, 50-Hz a.c. supply. If the current in the circuit is 1 A, determine (a) the impedance of the circuit, (b) the reactance of the capacitor, (c) the p.d. across the resistor and across the capacitor, (d) the power factor of the circuit and (e) the capacitance of the capacitor.
[(a) 250 Ω; (b) 229.1 Ω; (c) 100 V, 229.1 V; (d) 0.4; (e) 13.89 μF]

6.11 A resistor of resistance 30 Ω is connected in series with an inductor of inductance 0.1 H and a capacitor of capacitance 16.67 μF. The circuit is supplied at 200 V r.m.s., 159.2 Hz. Determine (a) the reactance of the inductor, (b) the reactance of the capacitor, (c) the circuit impedance, (d) the power factor of the circuit, (e) the current in the circuit, (f) the power consumed by the circuit and (g) the voltage across each element in the circuit. Draw the phasor diagram of the circuit to scale.
[(a) 100 Ω; (b) 60 Ω; (c) 50 Ω; (d) 0.6 (I lagging V); (e) 4 A; (f) 480 W; (g) V_L = 400 V, V_C = 240 V, V_R = 120 V]

6.12 Two circuits A and B are connected in series with one another to a 316.2 V r.m.s., 63.66 Hz supply. Circuit A contains $R = 10$ Ω and $L = 0.05$ H in series with one another; circuit B contains $R = 20$ Ω, $L = 0.1$ H and $C = 50$ μF in series with one another. Determine (a) the modulus of the impedance of each circuit, (b) the impedance of the circuits when connected in series with one another, (c) the voltage V_A across circuit A and V_B across circuit B, (d) the phase angle difference between V_A and V_B and (e) the power factor of the complete circuit.
[(a) 22.36 Ω; (b) 31.62 Ω; (c) 223.6 V; (d) V_A leads V_B by 90°; (e) 0.9487 (I lagging V)]

6.13 A resistor and a capacitor are connected in series with a resistanceless variable inductor. The circuit is energised by a 200-V, 50-Hz supply. When the inductance of the inductor is varied the circuit current has a maximum value of 0.314 A, at which time the voltage across the inductor is 300 V. Determine the value of the resistor and the capacitor in the circuit and also the value of the inductance at resonance. Draw to scale the phasor diagram of the circuit at resonance. Calculate also the Q-factor of the circuit at resonance.
[636.9 Ω; 3.33 μF; 3.04 H; 1.5]

6.14 A parallel circuit consists of a coil of resistance R and inductance L in series with a capacitor C. Deduce an expression for the resonant frequency of the circuit. Show that the impedance of the circuit at resonance is resistive and has the value L/CR Ω.

In the above circuit $R = 1$ Ω, $L = 10$ μH and $C = 10$ nF. Determine (a) the resonant frequency of the circuit, (b) the current drawn by the circuit at resonance if it is supplied at 10 V r.m.s. at the resonant frequency and (c) the Q-factor of the circuit.
[(a) 503 kHz; (b) 0.01 A; (c) 31.6]

6.15 An inductive load draws a current of 5 A at a power factor of 0.8 lagging from a 240-V, 50-Hz supply. To improve the power factor of the load, a 35-μF capacitor is connected to its terminals. Determine (a) the current drawn by the capacitor, (b) the total current drawn from the supply, (c) the new value of power factor and (d) the reactive volt-amperes consumed by the capacitor.
[(a) 2.64 A; (b) 4.02 A; (c) 0.996; (d) 633.6 VAr]

7 Complex Notation

7.1 Operator j

In the two preceding chapters the concept of the phase-angle difference between phasors was introduced, and in the following an analytical method is developed for the representation of phase difference. If we take phasor **OA** of length a in figure 7.1 and turn it through $90°$ in an *anticlockwise direction*, we say that we have *operated on* the phasor. To indicate that the phasor has rotated through $90°$, we write

$$\mathbf{OB} = \mathrm{j}a \qquad (7.1)$$

where j is the *$90°$ operator. Note* Length **OB** is shown in bold type to indicate that it is a phasor quantity.

If we operate once more on the phasor so that it assumes position **OC**, then

$$\mathbf{OC} = \mathrm{j}^2 a \qquad (7.2)$$

But, since **OC** = $-a$, then

$$\mathrm{j}^2 a = -a$$

or

$$\mathrm{j}^2 = -1 \qquad (7.3)$$

hence

$$\mathrm{j} = \sqrt{(-1)} \qquad (7.4)$$

Since the square root of minus unity does not exist as a 'real' quantity, the notion that quantities along the *quadrature axis* (that is, axis BOD) are *imaginary* has evolved. Consequently, the horizontal axis COA is sometimes described as the *real axis*. Readers will realise that simply to turn a quantity through $90°$ does not make it any more 'real' or 'imaginary' than it was in the first place. In the above context, the words 'real' and 'imaginary' merely refer to the horizontal and vertical directions, respectively.

Operating on phasor **OC** by j gives

$$\mathbf{OD} = \mathrm{j}^3 a \qquad (7.5)$$

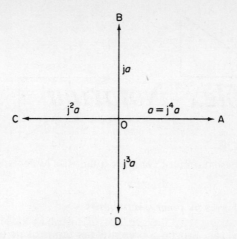

Figure 7.1 Operator j

and performing the operation once more yields

$$\mathbf{OA} = j^4 a \qquad (7.6)$$

Equation 7.6 yields

$$j^4 a = (j^2)^2 a = (-1)^2 a = a$$

Representation of phasors by Cartesian† or rectangular components

A complex quantity such as r_1 in figure 7.2a can be resolved into two components, each perpendicular to the other. For convenience, the horizontal (real) and perpendicular (imaginary) directions are chosen. Thus the phasor r_1 can be expressed in the form

$$r_1 = r_{h1} + j\, r_{v1} = r_1 \,(\cos\theta_1 + j\sin\theta) \qquad (7.7)$$

and, in figure 7.2b, phasor r_2 can be represented as

$$r_2 = r_{h2} - j\, r_{v2} = r_2 \,(\cos\theta_2 - j\sin\theta_2)$$

The *modulus* or magnitude of the phasor r_1 is calculated from the relationship

$$|r_1| = \sqrt{(r_{h1}{}^2 + r_{v1}{}^2)} \qquad (7.8)$$

and its *argument*‡ or phase angle is

$$\theta_1 = \tan^{-1}(r_{v1}/r_{h1}) \qquad (7.9)$$

†After Descartes the French mathematician and philospher.

‡Angle θ is sometimes written arg r_1.

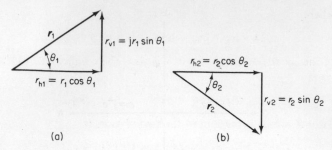

Figure 7.2   Representation of complex quantities by rectangular components

Representation of phasors by polar components

A complex quantity can also be represented in terms of its modulus r and phase angle θ, as follows

$$r_1 = r_1 \underline{/\theta_1}$$

In the case of figure 7.2b, r_2 is represented by

$$r_2 = r_2 \underline{/-\theta_2} \quad \text{or} \quad r_2 \overline{\backslash\theta_2}$$

7.2 Addition and Subtraction of Phasors Using Rectangular Components

It was shown in chapter 5 that the 'horizontal' and 'vertical' components of the sum of two phasors are determined by adding the respective components of the original phasors. Suppose that phasors X and Y in figure 7.3 are to be added

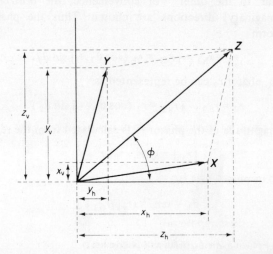

Figure 7.3   Addition of phasors by rectangular components

together, where

$$X = x_h + j x_v \quad \text{and} \quad Y = y_h + j y_v$$

Completing the parallelogram in figure 7.3 gives the components of the sum of the two phasors as

$$z_h = x_h + y_h \quad \text{and} \quad z_v = x_v + y_v$$

whence phasor Z, which is the sum of the two phasors is

$$Z = z_h + j z_v = (x_h + y_h) + j (x_v + y_v)$$

The modulus of Z is

$$|Z| = \sqrt{(z_h^2 + z_v^2)}$$

and

$$\phi = \tan^{-1} (z_v/z_h) = \cos^{-1} (z_h/|Z|)$$

Example 7.1

Voltages $V_1 = 10.3 + j2.1$ and $V_2 = 5.6 + j7.3$ are connected in series with each other. What is the magnitude and the phase angle of the resultant voltage V_S?

Solution

$$V_S = V_1 + V_2 = (10.3 + j2.1) + (5.6 + j7.3)$$
$$= (10.3 + 5.6) + j (2.1 + 7.3) = 15.9 + j9.4 \text{ V}$$

hence

$$|V_S| = \sqrt{(15.9^2 + 9.4^2)} = 18.47 \text{ V}$$

and

$$\phi = \tan^{-1} (9.4/15.9) = \tan^{-1} 0.5912 = 30.59°$$

Example 7.2

Determine the complex expression for the current I which is the phasor sum of the two currents $I_1 = -3 + j6$ and $I_2 = -1 - j8$. Evaluate the magnitude and the phase angle of I.

Solution

The phasor diagram is shown in figure 7.4.

$$I = I_1 + I_2 = (-3 + j6) + (-1 - j8)$$
$$= (-3 - 1) + j (6 - 8) = -4 - j2 \text{ A}$$

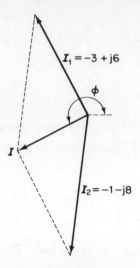

Figure 7.4

hence

$$|I| = \sqrt{[(-4)^2 + (-2)^2]} = 4.472 \text{ A}$$

and

$$\phi = \tan^{-1}(-2/-4) = 180° + 26.57° = 206.57° \text{ or } -153.43°$$

Phasor subtraction

In section 5.9 it was shown that the phasor difference is obtained by first adding a phase shift of $180°$ to the phasor to be subtracted and then performing phasor addition as follows.

The phasor difference $Z' = X - Y$ in figure 7.5 is obtained by adding $-Y$ to X, hence

$$Z' = X + (-Y) = (x_h + j\,x_v) - (y_h + j\,y_v)$$
$$= (x_h - y_h) + j\,(x_v - y_v)$$

Example 7.3

Determine the phasor difference $I = I_1 - I_2$, where (a) $I_1 = 10 + j8, I_2 = 5 + j3$, and (b) $I_1 = -3 + j8, I_2 = -8 - j\,3$.

Solution

(a) $$I = (10 + j8) - (5 + j3) = (10 - 5) + j(8 - 3)$$
$$= 5 + j5 \text{ A}$$

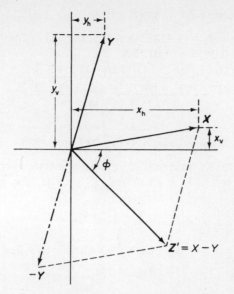

Figure 7.5 Subtracting phasors using rectangular components

(b) $\qquad I = (-3 + j8) - (-8 - j3) = (-3 + 8) + j(8 + 3)$
$\qquad\qquad\qquad = 5 + j11$

7.3 Addition and Subtraction of Phasors Using Polar Components

Adding and subtracting phasors by polar components is an inconvenient process, and is more easily performed by first converting the phasors into their rectangular components and completing the addition and subtraction as outlined above. The answer has then to be reconverted into polar co-ordinate form.

Example 7.4

Calculate (a) the phasor sum $I_1 + I_2$ and (b) the phasor difference $I_1 - I_2$ of the currents $I_1 = 2\underline{/45°}$ and $I_2 = 3\underline{/-60°}$.

Solution
$$I_1 = 2\underline{/45°} = 2(0.7071 + j0.7071) = 1.1412 + j1.1412 \text{ A}$$
$$I_2 = 3\underline{/-60°} = 3(0.5 - j0.866) = 1.5 - j2.598 \text{ A}$$

(a) If the phasor sum is I_s, then

$$I_s = I_1 + I_2 = (1.4142 + j1.4142) + (1.5 - j2.598)$$
$$= 2.9142 - j1.1838 \text{ A}$$
$$= \sqrt{(2.9142^2 + 1.1838^2)}\ \underline{/\tan^{-1}(-1.1838/2.9142)}$$
$$= 3.146\ \underline{/-22.1°}$$

(b) If the phasor difference is I_d, then

$$I_d = I_1 - I_2 = (1.4142 + j\,1.4142) - (1.5 - j2.598)$$
$$= -\,0.0858 + j4.0122 \text{ A}$$
$$= \sqrt{16.11} \;\underline{/\tan^{-1}\;[4.0142/(-0.0858)]}$$
$$= 4.013\;\underline{/180° - 88.78°} = 4.013\;\underline{/91.22°} \text{ A}$$

7.4 Multiplication of Phasors Using Rectangular Components

When multiplying phasors expressed in rectangular-component form the normal rules of algebra apply. If $V_1 = a + jb$ and $V_2 = c + jd$, the product of the two phasors is

$$V_1 \times V_2 = (a + jb) \cdot (c + jd)$$
$$= ac + jad + jbc + j^2\,bd = ac + j^2\,bd + j(ad + bc)$$

Since $j^2 = -1$, then

$$V_1 \times V_2 = (ac - bd) + j(ad + bc)$$

A product term of particular interest in electrical engineering is that of a pair of *conjugate complex numbers*. Two quantities are said to be conjugate if their magnitudes are equal to one another, but the phase angle of one phasor is $+\theta$ and that of the other is $-\theta$. Thus the complex conjugate of $(a + jb)$ is $(a - jb)$, and the product of a complex conjugate pair is

$$(a + jb) \cdot (a - jb) = a^2 + jab - jab - j^2 b = a^2 + b^2$$

Hence the product of a pair of complex conjugate numbers yields a product with no quadrature term, that is, no j term (see also section 7.6).

In the polar form, the complex conjugate of $V\underline{/\theta}$ is $V\underline{/-\theta}$, and the product of a complex conjugate pair is

$$V\underline{/\theta} \times V\underline{/-\theta} = V^2\,\underline{/\theta - \theta} = V^2\,\underline{/0°}$$

(see also section 7.5).

Example 7.5

Evaluate the product of the complex numbers $2 + j3$ and $4 - j5$.

Solution

$$(2 + j3) \cdot (4 - j5) = 8 - j10 + j12 - j^2 15$$
$$= [8 - (-15)] + j(-10 + 12) = 23 + j2$$

7.5 Multiplication of Phasors Using Polar Components

The rule that applies to the product of the phasors $R_1 = r_1 \underline{/\theta_1}$ and $R_2 = r_2 \underline{/\theta_2}$ is

$$R_1 \times R_2 = r_1 r_2 \underline{/\theta_1 + \theta_2}$$

Example 7.6

Determine the value of the product of the complex numbers $3.606 \underline{/56.3°}$ and $6.403 \underline{/-51.5°}$.

Solution

$$3.606 \underline{/56.3°} \times 6.403 \underline{/-51.5°}$$
$$= (3.606 \times 6.403) \underline{/56.3° + (-51.5°)}$$
$$= 23.09 \underline{/4.8°}$$

Note In practical terms, measurements of voltages and currents are usually obtained in polar form by means of a magnitude-measuring device such as a voltmeter or ammeter in conjunction with a phase-measuring device such as a cathode ray oscilloscope or a phasemeter.

7.6 Division of Phasors Using Rectangular Components

The quotient of two complex quantities can be obtained by first eliminating the quadrature terms in the denominator by a process known as *rationalising*. To carry this out, both the denominator and the numerator are multiplied by the complex conjugate of the denominator. If $V_1 = a + jb$ and $V_2 = c + jd$ then, in order to evaluate V_1/V_2 it is necessary to rationalise the expression as follows.

$$\frac{V_1}{V_2} = \frac{a + jb}{c + jd} = \frac{a + jb}{c + jd} \times \frac{c - jd}{c - jd} = \frac{ac + bd + j(bc - ad)}{c^2 + d^2}$$

$$= \frac{ac + bd}{c^2 + d^2} + j\frac{bc - ad}{c^2 + d^2}$$

Example 7.7

If two voltages $V_1 = 8.66 + j5$ and $V_2 = 10 - j10$ exist in an electrical circuit, determine the value of V_1/V_2.

Solution

$$\frac{V_1}{V_2} = \frac{8.66 + j5}{10 - j10} = \frac{(8.66 + j5)(10 + j10)}{(10 - j10)(10 + j10)}$$

$$= \frac{86.6 + j50 + j86.6 + j^2 50}{10^2 + 10^2} = \frac{36.6 + j136.6}{200}$$

$$= 0.183 + j0.683$$

7.7 Division of Phasors Using Polar Components

Division of complex numbers is carried out by first eliminating the quadrature part of the numerator, as in the case below using rectangular co-ordinates. If $V_1 = A\underline{/\theta_1}$ and $V_2 = B\underline{/\theta_2}$, then

$$\frac{V_1}{V_2} = \frac{A\underline{/\theta_1}}{B\underline{/\theta_2}} = \frac{A\underline{/\theta_1}}{B\underline{/\theta_2}} \times \frac{B\underline{/-\theta_2}}{B\underline{/-\theta_2}} = \frac{AB\underline{/\theta_1 - \theta_2}}{B^2\underline{/\theta_2 - \theta_2}}$$

$$= \frac{A}{B}\underline{/\theta_1 - \theta_2}$$

Example 7.8

Divide $10\underline{/30°}$ by $14.14\underline{/-45°}$, and give the result in polar form.

Solution

$$\frac{10\underline{/30°}}{14.14\underline{/-45°}} = \frac{10}{14.14}\underline{/30° - (-45°)} = 0.707\underline{/75°}$$

7.8 Representation of Voltage, Current, Impedance and Admittance in Complex Notation

In general terms, the voltage and current in an a.c. circuit are expressed in the form

$$V = V\underline{/\alpha} \quad \text{and} \quad I = I\underline{/\beta}$$

where α and β are the angular displacements of V and I, respectively, from a reference datum. It is often the case that either V or I is used as the reference phasor, so that either α or β has zero value.

Pure resistance

In a circuit containing only pure resistance, the current and voltage are in phase with one another so that

$$V = V\underline{/0°} \quad \text{and} \quad I = I\underline{/0°}$$

and the complex value of the circuit impedance Z is

$$Z = \frac{V}{I} = \frac{V}{I}\underline{/0°} = R \angle 0°$$

Pure inductance

In a circuit containing only pure inductance, the current lags behind the voltage by 90°, so that if

$$V = V\underline{/0°}$$

then

$$I = I\underline{/-90°} = -jI$$

The circuit impedance Z_L is

$$Z_L = \frac{V}{I} = \frac{V\underline{/0°}}{I\underline{/-90°}} = \frac{V}{I}\underline{/90°} = X_L\underline{/90°} = \omega L\underline{/90°} = jX_L = j\omega L$$

Pure capacitance

In this case the current leads the voltage across the capacitor terminals by 90°, hence if

$$V = V\underline{/0°}$$

then

$$I = I\underline{/90°}$$

and

$$Z_C = \frac{V}{I} = \frac{V\underline{/0°}}{I\underline{/90°}} = \frac{V}{I}\underline{/-90°} = X_C\underline{/-90°} = \frac{1}{\omega C}\underline{/-90°} = -\frac{j}{\omega C} = \frac{1}{j\omega C}$$

Series circuit impedance

In a series circuit the total impedance of the circuit is the sum of the individual impedances of the components; the individual impedances *must* be expressed in complex notation form, otherwise they cannot be added together. In the case of a circuit containing R, L and C in series, the circuit impedance is

$$Z = R + j\omega L + \frac{1}{j\omega C} = R + j\omega L - \frac{j}{\omega C} = R + j\left(\omega L - \frac{1}{\omega C}\right)$$

$$= R + j(X_L - X_C)$$

The modulus of the impedance is

$$|Z| = \sqrt{[R^2 + (X_L - X_C)^2]}$$

and the circuit phase angle is

$$\phi = \tan^{-1}\ [(X_L - X_C)/R]$$

and

$$\cos\phi = R/\ |Z|$$

If a number of complex impedances $Z_1, Z_2, \ldots Z_n$ are connected in series, then the effective complex impedance of the circuit is

$$Z_e = Z_1 + Z_2 + \ldots + Z_n$$

Impedance and admittance of parallel circuits

If two impedances Z_1 and Z_2 are connected in parallel with one another, the effective impedance Z_e of the combination is

$$Z_e = \frac{Z_1 Z_2}{Z_1 + Z_2}$$

The *admittance Y* of any circuit element is

$$Y = \frac{I}{V} = \frac{1}{Z} = G + jB \quad \text{siemens}$$

where G is the *conductance* of the element, and B is its *susceptance,* and is the quadrature part of the admittance.

For a purely resistive circuit

$$Y = \frac{1}{R} = G \quad \text{siemens}$$

For a circuit containing a pure inductive reactance

$$Y = \frac{1}{j\omega L} = \frac{-j}{X_L} = \frac{1}{X_L}\ \underline{/-90^\circ} = -jB \quad \text{siemens}$$

or

$$B = 1/X_L$$

For a circuit containing a pure capacitive reactance

$$Y = \frac{1}{1/j\omega C} = j\omega C = jX_C = X_C\underline{/90^\circ} = jB$$

or

$$B = X_C$$

For a circuit containing resistance, inductance and capacitance in parallel

$$Y = Y_1 + Y_2 + Y_3 = \frac{1}{R} - \frac{j}{X_L} + jX_C = \frac{1}{R} + j\left(X_C - \frac{1}{X_L}\right)$$

or

$$G = 1/R \quad \text{and} \quad B = \left(X_C - \frac{1}{X_L}\right)$$

7.9 Impedance and Admittance of Basic Circuits

In many instances it is convenient to be able to convert the impedance of, say, a series circuit into its admittance form. In the following the basic procedures for the conversions are outlined, and the results are given for the four basic circuits in figure 7.6.

Using the relationship

$$Z = 1/Y$$

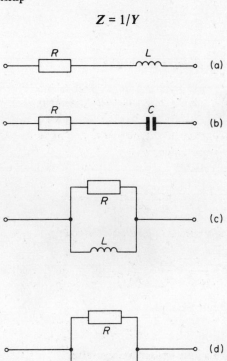

Figure 7.6 Basic *RL* and *RC* combinations

then, once either the admittance or the impedance is known, the other can be calculated. In the case of figure 7.6a, the circuit impedance is

$$Z = R + j\omega L$$

and its admittance is

$$Y = \frac{1}{Z} = \frac{1}{R + j\omega L} = \frac{R - j\omega L}{(R + j\omega L)(R - j\omega L)}$$

$$= \frac{R}{R^2 + (\omega L)^2} - j\frac{\omega L}{R^2 + (\omega L)^2} = G + jB$$

where

$$G = R/[R^2 + (\omega L)^2]$$

and

$$B = \omega L/[R^2 + (\omega L)^2]$$

Applying this technique to the four circuits in figure 7.6 yields the results in table 7.1.

Table 7.1

Circuit	Z	Y
figure 7.6a	$R + j\omega L$	$\dfrac{R}{R^2 + \omega^2 L^2} - j\dfrac{\omega L}{R^2 + \omega^2 L^2}$
figure 7.6b	$R - j/\omega C$	$\dfrac{R}{R^2 + 1/\omega^2 C^2} + \dfrac{j/\omega C}{R^2 + 1/\omega^2 C^2}$
figure 7.6c	$\dfrac{R\omega^2 L^2}{R^2 + \omega^2 L^2} + j\dfrac{\omega L R^2}{R^2 + \omega^2 L^2}$	$\dfrac{1}{R} - \dfrac{j}{\omega L}$
figure 7.6d	$\dfrac{R}{1 + \omega^2 C^2 R^2} - \dfrac{j\omega C R^2}{1 + \omega^2 C^2 R^2}$	$\dfrac{1}{R} + j\omega C$

7.10 Solved Series Circuit Examples

Example 7.9

A current of $(10 + j0)$ A flows through a circuit of impedance $(2 + j3)$ Ω. Calculate the value of the voltage across the resistive and reactive elements in the circuit, and also determine the value of the applied voltage. Compute also the phase difference between the applied voltage and the circuit current.

Solution

The circuit diagram is shown in figure 7.7a, together with its phasor diagram figure 7.7b. Now

$$V = IZ = (10 + j0)(2 + j3) = 20 + j30 = IR + jIX_L$$

Thus, the r.m.s. value of voltage across the resistive element is 20 V, and that across the reactive element is 30 V. The modulus of the supply voltage is

$$| V | = \sqrt{(20^2 + 30^2)} = 36.06 \text{ V}$$

and the circuit phase angle is

$$\phi = \tan^{-1}(V_L/V_R) = \tan^{-1}(30/20) = 56.31°$$

and, since the circuit has an inductive reactance, the current lags behind the applied voltage.

(a)

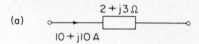

10 + j10 A

(b)

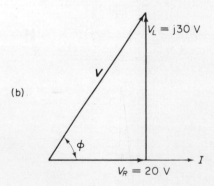

Figure 7.7

Example 7.10

A capacitance of 100 μF is connected in series with a coil of resistance 5 Ω and inductance 0.12 H, the combination being supplied by a 500 V r.m.s., 50 Hz source. Calculate (a) the r.m.s. value of the circuit current, (b) the phase angle and power factor of the circuit, (c) the r.m.s. voltage across the coil and (d) the r.m.s. value of voltage across the capacitor. Calculate also the power absorbed by the coil.

Solution

The circuit with its phasor diagram is shown in figure 7.8. The impedance Z_L of the coil is

$$Z_L = R + j\omega L = 5 + (j2\pi \times 50 \times 0.12) = 5 + j37.7 \ \Omega$$
$$= 38 \ \underline{/82.45^\circ} \ \Omega$$

The reactance of the capacitance is

$$Z_C = -j/\omega C = -j/(2\pi \times 50 \times 100 \times 10^{-6}) = -j31.83 \ \Omega$$

The circuit impedance Z is

$$Z = Z_L + Z_C = (5 + j37.7) + (-j31.83) = 5 + j5.87 \ \Omega$$
$$= 7.71 \ \underline{/49.58^\circ}$$

(a) $I = V/|Z| = 500/7.71 = 64.85$ A
(b) Phase angle $= \phi = 49.58^\circ$. Since $X_L > X_C$, the current lags behind the voltage.
(c) Voltage across $C = |IX_C| = 64.85 \times 31.83 = 2064$ V
(d) Voltage across the coil $= |IZ_L| = 64.85 \times 38 \ {=}\ 2464$ V

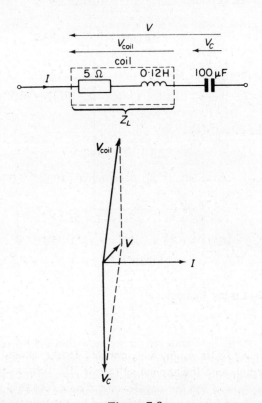

Figure 7.8

Note From the results in (c) and (d) above, the circuit operates at a frequency which is close to and just above the resonant frequency of the circuit.
   The power absorbed by the coil is

$$| I^2 R | = 64.85^2 \times 5 = 21\ 027\ W = 21.027\ kW$$

Example 7.11

When a voltage of $(50 + j75)$ V is applied to a circuit, the current is found to be $(2.25 + j6)$ A. Determine (a) the complex expression for the circuit impedance, (b) the phase angle between the current and the voltage (state if lagging or leading), and (c) the power consumed by the circuit.

Solution

$$V = 50 + j75 = 90.14\ \underline{/56.31^\circ}\ V$$
$$I = 2.25 + j6 = 6.41\ \underline{/69.44^\circ}\ A$$

(a) The circuit impedance is determined from the expression

$$Z = \frac{V}{I} = \frac{90.14\underline{/56.31^\circ}}{6.41\underline{/69.44^\circ}} = 14.06\underline{/56.31^\circ - 69.44^\circ}$$

$$= 14.06\underline{/-13.13^\circ}\ \Omega$$

$$= 14.06\ [\cos(-13.13^\circ) + j\sin(-13.13^\circ)] = 13.7 - j3.2\ \Omega$$

(b) From the results of part (a) above

$$\phi = 13.13^\circ\ (I\ leading\ V)$$

Note The phase angle could have been determined from the phase angles of the voltage and current as follows

$$\phi = 56.31^\circ - 69.44^\circ = -13.13^\circ$$

(c) Power $= VI \cos\phi = 90.14 \times 6.41 \times \cos(-13.13^\circ) = 562.7\ W$

7.11 Solved Parallel Circuit Examples

Example 7.12

When a 250 V r.m.s., 50 Hz supply is connected to the circuit in figure 7.9, the phase angle between I_B and the applied voltage is 30°. Determine (a) the complex expressions for I_A, I_B and I in rectangular co-ordinate form, (b) the r.m.s. value of each of the currents, and (c) the power consumed by the inductive branch.

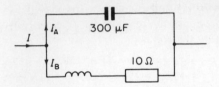

Figure 7.9   Circuit for example 7.12

Solution

(a) The impedance triangle for the inductive branch is shown in figure 7.10a, in which

$$Z_B = R/\cos 30° = 10/0.866 = 11.55 \ \Omega$$

therefore

$$Z_B = 11.55 \ \underline{/30°}$$

Hence

$$I_B = V/Z_B = 250/11.55 \ \underline{/30°} = 21.65 \ \underline{/-30°} \ A$$
$$= 18.75 - j10.83 \ A$$

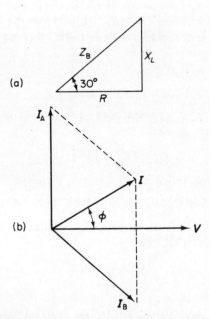

Figure 7.10   (a) Impedance triangle for the inductive branch of figure 7.9, and
(b) the phasor diagram for example 7.12

The impedance of the capacitor is

$$Z_C = -j/2\pi fC = -j/2\pi \times 50 \times 300 \times 10^{-6} = -j10.61 \ \Omega$$

Hence

$$I_A = V/X_C = 250/(-j10.61) = j23.56 \ A$$

and

$$I = I_A + I_B = j23.56 + (18.75 - j10.83)$$
$$= 18.75 + j12.73A$$

(b) From the above

$$|I_A| = 23.56 \ A$$
$$|I_B| = 21.65 \ A$$
$$|I| = \sqrt{(18.75^2 + 12.73^2)} = 22.66 \ A$$

(c)   Power consumed by the inductor $= I_B{}^2R = 21.66^2 \times 10$

$$= 4687 \ W$$

Example 7.13

Impedances $Z_1 = (20 + j15) \ \Omega$ and $Z_2 = (10 - j60) \ \Omega$ are connected in parallel with one another. If the supply frequency is 50 Hz, determine the resistance and the inductance or capacitance in each branch. Calculate also the complex admittance of the parallel circuit and the phase angle between the resultant current and the applied voltage.

Solution

Since $Z_1 = (20 + j15) \ \Omega$, it consists of a resistance of 20 Ω in series with an inductive reactance of j15 Ω. Hence

$$2\pi fL = 15 \ \Omega$$

or

$$L = 15/2\pi f = 15/(2\pi \times 50) = 0.04774 \ H = 47.74 \ mH$$

Also, since $Z_2 = (10 - j60) \ \Omega$, it consists of a resistance of 10 Ω in series with a capacitance of reactance −j60 Ω that is

$$1/2\pi fC = 60 \ \Omega$$

or

$$C = 1/(2\pi f \times 60) = 1/(2\pi \times 50 \times 60) = 0.000053 \ F$$
$$= 53 \ \mu F$$

Now

$$Y_1 = \frac{1}{Z_1} = \frac{1}{20 + j15} = \frac{20 - j15}{20^2 + 15^2} = \frac{20 - j15}{625}$$

$$= 0.032 - j0.024 \text{ S}$$

and

$$Y_2 = \frac{1}{Z_2} = \frac{1}{10 - j60} = \frac{10 + j60}{10^2 + 60^2} = \frac{10 + j60}{3700}$$

$$= 0.0027 + j0.0162 \text{ S}$$

The total admittance of the parallel circuit is

$$Y = Y_1 + Y_2 = 0.032 - j0.024 + 0.0027 + j0.0162$$

$$= 0.0347 - j0.0078 \text{ S}$$

The phasor diagram showing the applied voltage and resultant current is shown in figure 7.11. From the figure

$$\phi = \tan^{-1} \frac{VB}{VG} = \tan^{-1} \frac{B}{G} = \tan^{-1} \frac{0.0078}{0.0347} = \tan^{-1} 0.2248$$

$$= 12.67° \ (I \text{ lagging } V)$$

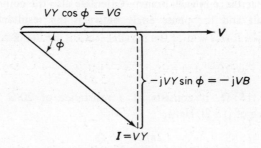

Figure 7.11   Phasor diagram for example 7.13

7.12 Solved Series–Parallel Example

Suppose in the circuit in figure 7.12 we need to evaluate the impedance of Z and to determine the elements it contains given that the supply voltage is $100\underline{/0°}$ V and that $I = 10\underline{/0°}$ A. The value of the voltage V_P across the parallel circuit is also required.

If the impedance of the complete circuit is Z_1, then

$$Z_1 = V/I = 100\underline{/0°} \ /10\underline{/0°} = 10\underline{/0°} \ \Omega = 10 + j0 \ \Omega$$

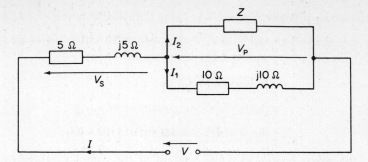

Figure 7.12 Circuit diagram for the problem in section 7.12

The voltage V_S across the series section of the circuit is

$$V_S = I(5 + j5) = 10(5 + j5) = 50 + j50 \text{ V}$$

Now

$$V = V_P + V_S$$

hence

$$V_P = V - V_S = (100 + j0) - (50 + j50) = 50 - j50 \text{ V}$$

$$= 70.7\underline{/-45°} \text{ V}$$

The current I_1 in the lower arm of the parallel section is calculated as follows

$$I_1 = V_P/(10 + j10) = 70.7\underline{/-45°}/14.14\underline{/45°} = 5\underline{/-90°} \text{ A}$$

$$= -j5 \text{ A}$$

Also, since $I = I_1 + I_2$, then

$$I_2 = I - I_1 = (10 + j0) - (-j5) = 10 + j5 \text{ A}$$

$$= 11.18\underline{/26.57°} \text{ A}$$

The value of impedance Z is calculated as follows

$$Z = V_P/I_2 = 70.7\underline{/-45°}/11.18\underline{/26.57°} = 6.32\underline{/-71.57°}$$

$$= 6.32\,[\cos(-71.57°) + j\sin(-71.57°)]$$

$$= 6.32\,(0.3162 - j0.9487) = 2 - j6 \text{ }\Omega$$

That is, Z consists of a resistor of 2 Ω resistance in series with a capacitor whose reactance is 6 Ω at the operating supply frequency.

7.13 Calculation of Power, VA and VAr Using Complex Notation

Suppose that the voltage and current phasors associated with a circuit are as shown in figure 7.13. The power consumed by the circuit is

$$P = VI \cos \phi = VI \cos (\alpha - \beta)$$
$$= \text{the in-phase component of } VI\underline{/\alpha - \beta}$$
$$= \text{the in-phase component of } (V\underline{/\alpha} \times I\underline{/-\beta})$$

That is, **the power consumed is the product of the complex expressions of the voltage and the** *conjugate* **of the current.**

If $V = a + jb$ and $I = c + jd$, the complex conjugate I^* is $I^* = c - jd$. The power consumed is given by the 'real' part of the product VI^* as follows

$$VI^* = S = (a + jb)(c - jd) = (ac + bd) + j(bc - ad)$$
$$= P + jQ = \text{power} + j \text{ (reactive volt amperes)}$$

where

$$\text{Power} = P = ac + bd \quad \text{watts}$$
$$\text{Reactive VA} = Q = bc - ad \quad \text{VAr}$$

and

$$\text{Apparent power} = S = \sqrt{(P^2 + Q^2)} = \sqrt{[(ac + bd)^2 + (bc - ad)^2]}$$

hence

$$S = P + jQ = S\underline{/\phi} \quad \text{VA}$$

Example 7.14

A voltage of $(100 + j0)$ V is applied to an inductive circuit of impedance $(1.732 + j1)\ \Omega$. Calculate the power consumed by the circuit and also the VA and VAr consumption. What is the power factor of the circuit?

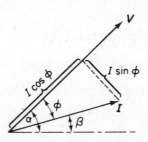

Figure 7.13 Calculation of power, VA and VAr in a.c. circuits

Solution

$$Z = 1.732 + \text{j}1 \ \Omega = 2\underline{/30^\circ} \ \Omega$$

Hence

$$I = 100/(2\underline{/30^\circ}) = 50\underline{/-30^\circ} \ \text{A}$$

The complex conjugate I^* of the current is

$$I^* = 50\underline{/30^\circ}$$

Hence

$$VI^* = 100\underline{/0^\circ} \times 50\underline{/30^\circ} = 5000\underline{/30^\circ} \quad \text{VA}$$

$$= 5000(0.866 + \text{j}0.5) = 4330 + \text{j}2500 \quad \text{VA}$$

From the above expression

Apparent power = S = 5000 VA
Power = P = 4330 W
Reactive VA = Q = 2500 VAr

and

Power factor = $\cos \phi = \cos 30^\circ = 0.866$

7.14 Operator h (or a)

In three-phase work the supply voltage phasors are displaced by 120° from one another — see figure 7.14 in which the voltage phasors are V_R, V_Y and V_B. In three-phase analysis it is sometimes convenient to use an operator which causes the phasors to rotate through 120° rather than 90°. Thus in figure 7.14

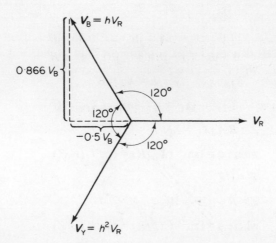

Figure 7.14 Operator h

$$V_B = V_R \underline{/120^\circ} = hV_R$$

where h is an operator having the value $1\underline{/120^\circ}$; hence

$$h = 1\underline{/120^\circ} = -0.5 + j0.866$$

Operator h is sometimes described as operator a. Also, from figure 7.14

$$V_Y = V_R \underline{/240^\circ} = h^2 V_R$$

where

$$h^2 = 1\underline{/240^\circ} = -0.5 - j0.866$$

and

$$V_R = V_R \underline{/360^\circ} = h^3 V_R$$

Hence

$$h^3 = 1\underline{/360^\circ} = 1 + j0$$

Summary of essential formulae

$$j^2 = -1 = 1\underline{/180^\circ}$$

$$j = 1\underline{/90^\circ} = \sqrt{(-1)}$$

$$I = I\underline{/\theta} = I\,(\cos\theta + j\sin\theta) = A + jB$$

$$|I| = \sqrt{(A^2 + B^2)} \quad \text{and} \quad \theta = \tan^{-1}(B/A) = \cos^{-1}(A/I)$$

$$I\underline{/\phi_1} \times Z\underline{/\phi_2} = IZ\underline{/\phi_1 + \phi_2}$$

$$\frac{V\underline{/\phi_1}}{I\underline{/\phi_2}} = \frac{V}{I}\underline{/\phi_1 - \phi_2}$$

$$\frac{1}{A + jB} = \frac{A - jB}{(A + jB)(A - jB)} = \frac{A - jB}{A^2 + B^2}$$

Inductive reactance: $\quad X_L = jX_L = j2\pi fL = X_L\underline{/90^\circ}$

Capacitive reactance: $\quad X_C = -jX_C = -j/2\pi fC = X_C\underline{/-90^\circ}$

Series RL circuit: $\quad Z = R + jX_L = Z\underline{/\phi}$

   where $\phi = \tan^{-1}(X_L/R) = \cos^{-1}(R/Z)$

   $Y = 1/Z$

Series RC circuit: $\quad Z = R - jX_C = Z\underline{/-\phi}$

   where $\phi = \tan^{-1'}(X_C/R) = \tan^{-1}(R/Z)$

   $Y = 1/Z$

Impedances in series: $\quad Z = Z_1 + Z_2 = (R_1 + jX_1) + (R_2 + jX_2)$

$$= (R_1 + R_2) + j(X_1 + X_2) = Z\underline{/\phi}$$

where $Z = \sqrt{[(R_1 + R_2)^2 + (X_1 + X_2)^2]}$

and $\quad \phi = \tan^{-1}[(X_1 + X_2)/(R_1 + R_2)]$

Admittances in parallel: $\quad Y = Y_1 + Y_2 = (G_1 + jB_1)(G_2 + jB_2)$

$$= (G_1 + G_2) + j(B_1 + B_2) = Y\underline{/\phi}$$

where $Y = \sqrt{[(G_1 + G_2)^2 + (B_1 + B_2)^2]}$

and $\quad \phi = \tan^{-1}[(B_1 + B_2)/(G_1 + G_2)]$

Apparent power: $\quad S = VI^* = P + jQ \quad$ volt amperes

where I^* is the complex conjugate of I

P is the power in watts

Q is the reactive power in VAr

Operator h (or a): $\quad h = 1\underline{/120°} = -0.5 + j0.866$

PROBLEMS

7.1 The following complex voltages are connected in series with one another; determine the expression for the resultant voltage (a) in rectangular coordinate form, (b) in polar coordinate form: $6 + j7$ V; $8 - j5$ V; $-9 - j8$ V; $-3 + j2$ V.
[(a) $2 - j4$ V; (b) $4.472\underline{/-63.44°}$]

7.2 Two electrical circuits X and Y are connected in parallel with one another, circuit X drawing a current of $2 + j3$ A. If the current taken by the parallel circuit is $8.25\underline{/-14.04°}$ A, determine the current in circuit Y (a) in rectangular form, (b) in polar form.
[(a) $6 - j5$ A; (b) $7.81\underline{/-39.81°}$ A]

7.3 If a current of $5 + j6$ A flows in a circuit of impedance $10\underline{/30°}$ Ω, determine the voltage applied to the circuit; carry the calculation out completely in (a) rectangular coordinate form, (b) polar coordinate form.
[(a) $13.31 + j76.96$ V; (b) $78.1\underline{/80.19°}$ V]

7.4 A voltage of $240\underline{/120°}$ V is applied to a circuit of impedance $10 + j5.77$ Ω. Calculate the current in the circuit using (a) polar coordinate values, (b) rectangular coordinate values.
[(a) $20.78\underline{/90°}$ A; (b) $j20.78$ A]

7.5 A resistanceless coil of inductance 0.1 H and a resistor of 500 Ω are connected (a) in series, (b) in parallel with a 10-V, 1-kHz a.c. supply. Determine expressions for (i) the impedance and admittance of the circuits and (ii) the current in the circuits in both rectangular and polar form.

[(a) (i) 500 + j 628.3 Ω, (ii) 803 $\underline{/51.49°}$ Ω; (b) (i) 0.0078 − j0.01 A, (ii) 0.0125 $\underline{/-51.49°}$ A]

7.6 The p.d. across a circuit is given by 25 + j40 V. If the circuit consists of a coil of resistance 12.5 Ω and inductance 0.04 H, and the angular frequency of the supply is 500 rad/s, determine complex expressions for the current (a) in polar form, (b) in rectangular form.

[(a) 2 $\underline{/0°}$ A; (b) 2 + j0 A]

7.7 The p.d. across a circuit is given by (3000 + j600) V and the current through it by (10 − j5) A. Determine the power, the reactive VA and the apparent power consumed. State whether the reactive VA is lagging or leading.

[27 kW; 21 kVAr (lagging); 34.2 kVA]

8 Polyphase Alternating Current Circuits

8.1 Introduction to Polyphase Systems

A polyphase system is one which comprises two or more sets of supply voltages which have a fixed phase angle difference between them. In the two-phase system (figure 8.1a) the phase difference is $90°$, and in the three-phase system (figure 8.1b) it is $120°$. Although the three-phase system is the most popular, some installations use six-, twelve- and twenty-four-phase systems.

A feature of the three-phase system is that, for a given amount of power transmitted through the system, the polyphase supply requires a smaller volume of copper (that is, conductor material) than does the equivalent single-phase system.

8.2 Generating Three-phase E.M.F.s

In three-phase work it is necessary to identify each of the phase voltages, and a convention which has been adopted is to call the phases the R (red), Y (yellow) and B (blue) phases, respectively. In another convention they are known as the A (or a), B (or b) and C (or c) phases.

A single-phase supply is generated by a single loop or coil of wire that rotates in a magnetic field in the manner shown in figure 8.2a. At the instant of time shown, conductors R and R′ move in a direction which is parallel to the magnetic field and do not cut the magnetic flux, and the instantaneous value, $v_{RR'}$, of the e.m.f. induced in the coil, is zero. As the loop rotates in an anticlockwise direction, conductor R passes under the upper pole piece when the polarity of the instantaneous e.m.f. induced in the coil is positive, having a maximum value of V_m when the coil is perpendicular. When the angle turned through exceeds $180°$, conductor R passes under the lower pole piece and the polarity of the induced e.m.f. becomes negative.

In the case of the polyphase generator, figure 8.2b, the e.m.f.s are induced in coils that are physically displaced from one another by $120°$. The voltage induced in winding RR′ is as described for the single-phase case. At the instant of time considered in the figure, conductor Y is under the lower pole piece and conductor B is under the upper pole piece. Consequently, the polarities of the induced e.m.f.s $v_{YY'}$ and $v_{BB'}$ are negative and positive respectively. As the conductors rotate,

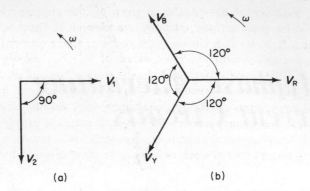

Figure 8.1   (a) Two-phase and (b) three-phase supply systems

conductor Y passes under the centre of the lower pole, when the maximum value of
e.m.f. with negative polarity is induced in it, before it moves towards the position
assumed initially by conductor R, which it reaches after 120° of rotation. Thus the
waveform associated with voltage $v_{YY'}$ is generated by phasor Y in figure 8.2b,
which *lags* behind phasor R by 120°. Conductor B has to rotate through 240°
before it reaches the initial position of conductor R, so that voltage $v_{BB'}$ is
represented by phasor B, which *lags* behind phasor R by 240°. Alternatively, phasor
B can be regarded as leading phasor R by 120°.

As in the single-phase case, the phasors associated with three-phase systems are
scaled to represent r.m.s. values rather than maximum values.

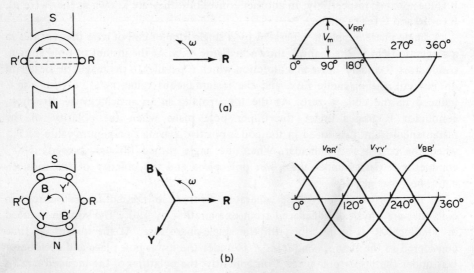

Figure 8.2   Generating (a) a single-phase supply and (b) a three-phase supply

The *phase sequence* of the phasors is given by the sequence in which the conductors pass the point initially taken by conductor R in figure 8.2b. Thus the phase sequence is **R, Y, B** and is known as the *positive phase sequence* (P.P.S.).

In all normal supply systems the r.m.s. values of V_R, V_Y, and V_B are equal to one another, and the phase difference between pairs of voltage phasors is 120°. Such systems are known as *symmetrical supply systems*. In a few cases the magnitudes of the voltages differ from one another and also the phase difference between pairs of phasors may not be 120°; in such cases the supply is said to be *unsymmetrical*. Unsymmetrical conditions can occur as a result of a fault, for example a short-circuit between a pair of lines, or an open-circuited line.

8.3 Star Connection of Three-phase Windings

If ends R′, Y′, and B′ of the coils in figure 8.2b are joined together, the resulting connection is known as the *star connection*, shown in figure 8.3a. The common connection of the three windings is known as the *neutral point*, N, of the system; the voltages between terminals R–N, Y–N and B–N are the *phase voltages* of the system and, in a symmetrical system, are equal in magnitude. The symbol given to the phase voltage is V_P. Since terminals R, Y, and B are connected to the outgoing lines, they are known as the *line terminals*, and the voltages between them are the *line-to-line voltages* (or *line voltage*), V_L.

The phasor diagram for the system voltages is shown in figure 8.3b, in which the phase voltage between the neutral point and line R is V_{RN}, the phase voltage between the neutral and line Y is V_{YN}, and that between the neutral and line B is V_{BN}, where

V_{RN} = voltage of line R with respect to the neutral point
V_{YN} = voltage of line Y with respect to the neutral point
V_{BN} = voltage of line B with respect to the neutral point

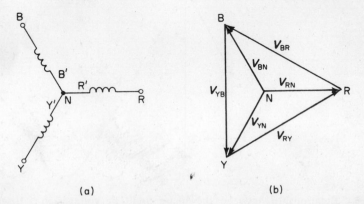

(a)   (b)

Figure 8.3   Phase voltages and line voltages in balanced three-phase system

(The double subscript notation used here corresponds to that developed in chapter 1.) The voltage between lines R and Y is V_{RY}, where

$$V_{RY} = \text{voltage of line R relative to line Y} = V_{RN} - V_{YN}$$

also

$$V_{BR} = \text{voltage of line B relative to line R} = V_{BN} - V_{RN}$$

and

$$V_{YB} = \text{voltage of line Y relative to line B} = V_{YN} - V_{BN}$$

The way in which the line voltages are determined from the above equations is shown in figure 8.4. If V_P is the r.m.s. value of each phase voltage, then

$$V_{RN} = V_P\underline{/0^\circ} = V_P(1 + j0) \tag{8.1}$$

$$V_{YN} = V_P\underline{/240^\circ} = V_P(-0.5 - j0.866) = h^2 V_P \tag{8.2}$$

$$V_{BN} = V_P\underline{/120^\circ} = V_P(-0.5 + j0.866) = h V_P \tag{8.3}$$

where h is the complex operator $1\underline{/120^\circ}$ (see section 7.14). From the equations given earlier

$$V_{RY} = V_{RN} - V_{YN} = V_P(1 + j0) - V_P(-0.5 - j0.866)$$

$$= V_P(1.5 + j0.866) = \sqrt{3}V_P\underline{/30^\circ} \tag{8.4}$$

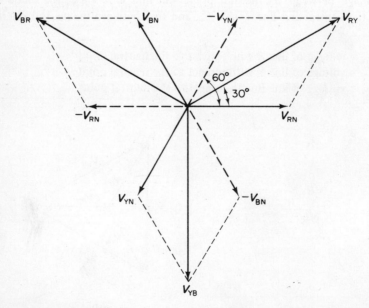

Figure 8.4   Relationship between the phase and line voltages in a three-phase system

From equation 8.4, the magnitude of V_{RY} (that is, the line voltage) is $\sqrt{3}V_P$; hence

Line voltage magnitude = $\sqrt{3}$ × Phase voltage magnitude

Thus, a three-phase star-connected supply with a phase voltage of 240 V has a line voltage of 416 V.

8.4 Star-connected Loads

A *three-phase four-wire* supply system is illustrated in figure 8.5. The load is supplied via lines R, Y and B, and the neutral point, N, of the supply is linked to the star point, S, of the load by a link known as the *neutral wire*. The load is described as a *balanced load* if the impedance and the power factor are the same in each phase of the load. If this is not the case, then the load is said to be *unbalanced*.

In the case of the star-connected system, the current flowing in the phase winding (or load) is equal to the current flowing in the line. That is

Phase current = Line current

or

$$I_P = I_L$$

The relationship between the neutral wire current and the currents in the loads (the phase currents) is obtained by applying Kirchhoff's first law to the star point of the load as follows

$$I_N = I_R + I_Y + I_B \qquad (8.5)$$

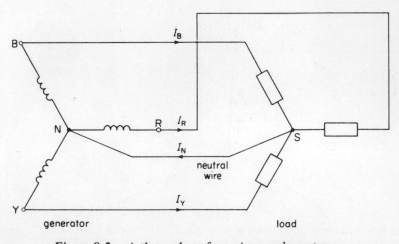

Figure 8.5   A three-phase four-wire supply system

Balanced loads

If the impedance of each load is $Z\underline{/\phi}$, that is, an inductive load, the phase currents are

$$I_R = V_{RN}/Z\underline{/\phi} = V_P\underline{/0^\circ}/Z\underline{/\phi} = \frac{V_P}{Z}\underline{/-\phi}$$

$$I_Y = V_{YN}/Z\underline{/\phi} = V_P\underline{/240^\circ}/Z\underline{/\phi} = \frac{V_P}{Z}\underline{/240^\circ - \phi} \qquad\qquad (8.6)$$

$$I_B = V_{BN}/Z\underline{/\phi} = V_P\underline{/120^\circ}/Z\underline{/\phi} = \frac{V_P}{Z}\underline{/120^\circ - \phi}$$

Hence

$$I_N = I_R + I_Y + I_B = \frac{V_P}{Z}(1\underline{/-\phi} + 1\underline{/240^\circ - \phi} + 1\underline{/120^\circ - \phi}) \qquad (8.7)$$

$$= 0$$

The terms within the brackets in equation 8.7 are shown in figure 8.6, and their phasor sum is seen to be zero. That is, **in a balanced three-phase star-connected load the instantaneous value of the neutral current is zero.** The reason for this can be physically argued from equation 8.6 as follows. If the load in each phase is a pure resistance of 100 Ω and if the phase voltage is 100 V, then

$$I_R = 100\underline{/0^\circ}/100 = 1\underline{/0^\circ} = 1 + j0 \text{ A}$$

$$I_Y = 100\underline{/240^\circ}/100 = 1\underline{/240^\circ} = -0.5 - j0.866 \text{ A} \qquad (8.8)$$

$$I_B = 100\underline{/120^\circ}/100 = 1\underline{/120^\circ} = -0.5 + j0.866 \text{ A}$$

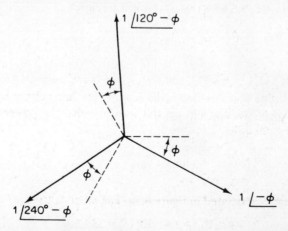

Figure 8.6   Phasor diagram corresponding to equation 8.6

Hence the magnitude of the current in each line is the same, but the *direction* of flow of the 'real' and quadrature components differs. From equation 8.8, line R carries a 'real' component of current of +1 A, that is, 1 A *towards* the load, while lines Y and B each carry a 'real' component of current of 0.5 A *away* from the load. Similarly, the quadrature component of the Y line current is balanced by the quadrature component of the B line current. Consequently, no current flows in the neutral wire under conditions of balanced loading.

This fact is utilised in many industrial situations in which motors present a balanced load to the supply, and many installations use a *three-phase three-wire supply system* which does not employ a neutral wire.

When designing an electrical installation, it is necessary to know the characteristics of the connected loads, described as follows. The analysis of the current waveform of a fluorescent lamp shows it to have a large third-harmonic component, in some cases as much as one-third of the total current. When identical fluorescent lamps are connected in the form of a star load to a three-phase supply, the fundamental (mains) frequency components of current cancel out in the neutral wire in the manner described above. However, the phase relationships between the harmonic components are such that the harmonics add together. Thus the neutral wire current from this kind of load consists entirely of third-harmonic current (that is, three times the supply frequency) and is of large value (nearly equal in magnitude to the line current), even though the load is nominally balanced. Moreover, power factor correction of the lamp does not improve the situation since it simply reduces the fundamental frequency (mains frequency) component of current, and does not reduce the third-harmonic current to the same extent.

Unbalanced loads, three-phase four-wire system

In the case of an unbalanced load the sum of the line currents is not zero, and a current flows in the neutral wire (see example 8.1 below).

Example 8.1

Loads of $(10 + j15)$ Ω, $(25 + j0)$ Ω, and $(30 - j50)$ Ω are connected in the R, Y, and B phases, respectively, of a star-connected load. If the line voltage is 550 V r.m.s., calculate the values of the line and neutral wire currents, the power consumed by each load and the total power consumed.

Solution

The circuit diagram is illustrated in figure 8.7a. The phase voltage is

$$V_P = V_L/\sqrt{3} = 550/\sqrt{3} = 317.5 \text{ V}$$

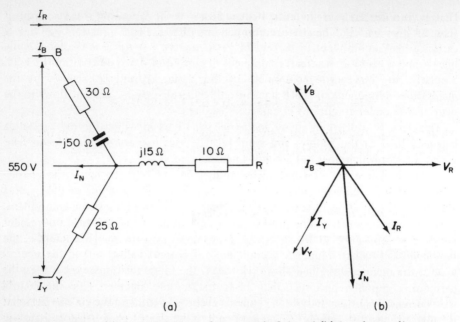

Figure 8.7   (a) Circuit diagram for example 8.1 and (b) its phasor diagram

Now, $Z_R = (10 + j15)\ \Omega, Z_Y = (25 + j0)\ \Omega$, and $Z_B = (30 - j50)\ \Omega$ hence

$$I_R = \frac{317.5 + j0}{10 + j15} = \frac{317.5(10 - j15)}{10^2 + 15^2}$$

$$= 9.77 - j14.65\ \text{A} = 17.61\underline{/56.3^\circ} \quad \text{(lagging)} \quad \text{A}$$

$$I_Y = \frac{317.5(-0.5 - j0.866)}{25 + j0}$$

$$= -6.35 - j11\ \text{A} = 12.7\underline{/-120^\circ} \quad \text{(that is, in phase with } V_Y) \quad \text{A}$$

$$I_B = \frac{317.5(-0.5 + j0.866)}{30 - j50} = -5.44 + j0.092 \quad \text{A}$$

$$= 5.45\underline{/179^\circ} \quad \text{(that is, leading } V_B \text{ by } 59^\circ) \quad \text{A}$$

Note The currents and their phase angles can also be evaluated in polar form directly by using the voltage and impedance values in polar form.

$$I_N = I_R + I_Y + I_B = (9.77 - j14.65) + (-6.35 - j11) + (-5.44 + j0.092)$$

$$= -2.02 - j25.56\ \text{A} = 25.63\underline{/-94.52^\circ} \quad \text{A}$$

That is, the neutral current leads V_Y by $25.48°$ or it can be regarded as lagging behind V_R by $94.52°$. The power consumed in each phase is

$$P_R = I_R{}^2 R_R = 17.61^2 \times 10 = 3101 \text{ W} = 3.101 \text{ kW}$$

$$P_Y = I_Y{}^2 R_Y = 12.7^2 \times 25 = 4032 \text{ W} = 4.032 \text{ kW}$$

$$P_B = I_B{}^2 R_B = 5.45^2 \times 30 = 891 \text{ W} = 0.891 \text{ kW}$$

where R_R, R_Y, and R_B are the resistive components of the impedances in the R, Y, and B phases respectively. The total power consumed is

$$P = P_R + P_Y + P_B = 8.024 \text{ kW}$$

Unbalanced loads, three-phase three-wire systems

In the case of a three-phase three-wire system, the neutral wire is omitted – see figure 8.8. Thus

$$I_R + I_Y + I_B = 0$$

If the load is balanced the sum of the line currents is zero, and the voltage between the star and neutral points, V_{SN}, is also zero. In the case of an unbalanced load the sum of the line currents must again be zero; the unbalance in the system loading usually manifests itself as a shift in the star point voltage, so that V_{SN} is no longer zero (see example 8.2 below). The value of V_{SN} can be calculated using Millman's theorem (see section 2.12) which, when applied to figure 8.8 gives

$$V_{SN} = \frac{V_R Y_R + V_Y Y_Y + V_B Y_B}{Y_R + Y_Y + Y_B} \tag{8.9}$$

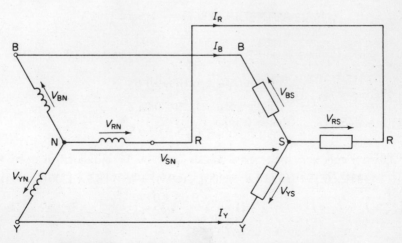

Figure 8.8   A three-phase three-wire supply system

where $Y_R = 1/Z_R$, $Y_Y = 1/Z_Y$, and $Y_B = 1/Z_B$. Also

$$\left.\begin{array}{l} V_{RN} = V_{RS} + V_{SN} \\ V_{YN} = V_{YS} + V_{SN} \\ V_{BN} = V_{BS} + V_{SN} \end{array}\right\} \tag{8.10}$$

and

$$\left.\begin{array}{l} I_R = V_{RS}/Z_R = V_{RS}Y_R \\ I_Y = V_{YS}/Z_Y = V_{YS}Y_Y \\ I_B = V_{BS}/Z_B = V_{BS}Y_B \end{array}\right\} \tag{8.11}$$

Example 8.2

Three non-reactive resistors of 10 Ω, 20 Ω and 25 Ω are connected in the R, Y, and B lines, respectively, of a star-connected load. The load is energised by a 400-V r.m.s. symmetrical three-phase supply. Calculate (a) the voltage between the star point of the load and the neutral point of the supply, (b) the voltage across each load, (c) the current in each line, and (d) the power consumed by each resistor.

Solution

$$V_P = V_L/\sqrt{3} = 400/\sqrt{3} = 231 \text{ V}$$

$$V_{RN} = 231 + j0$$

$$V_{YN} = 231(-0.5 - j0.866) = -115.5 - j200 \text{ V}$$

$$V_{BN} = 231(-0.5 + j0.866) = -115.5 + j200 \text{ V}$$

and

$$Y_R = 1/Z_R = 1/10 = 0.1 \text{ S}$$

$$Y_Y = 1/Z_Y = 1/20 = 0.05 \text{ S}$$

$$Y_B = 1/Z_B = 1/25 = 0.04 \text{ S}$$

From equation 8.9

(a) $V_{SN} = \dfrac{(231 \times 0.1) + [(-115.5 - j200) \times 0.05)] + [(-115.5 + j200) \times 0.04)]}{0.1 + 0.05 + 0.04}$

$$= 66.87 - j10.53 \text{ V} = 67.69\underline{/-8.95^\circ} \text{ V}$$

The phasor diagram of the system is shown in figure 8.9.

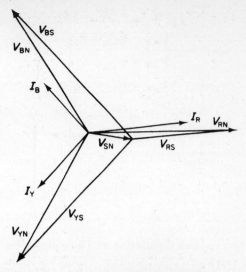

Figure 8.9 Phasor diagram for example 8.2

(b) From equation 8.10

$$V_{RS} = V_{RN} - V_{SN} = (231 + j0) - (66.87 - j10.53)$$

$$= 164.13 + j10.53 = 164.5\underline{/3.67^\circ} \text{ V}$$

$$V_{YS} = V_{YN} - V_{SN} = (-115.5 - j200) - (66.87 - j10.53)$$

$$= -182.37 - j189.47 = 263\underline{/-133.91^\circ}$$

$$V_{BS} = V_{BN} - V_{SN} = (-115.5 + j200) - (66.87 - j10.53)$$

$$= -182.37 + j210.53 = 278.5\underline{/130.9^\circ} \text{ V}$$

(c) From equation 8.11

$$I_R = V_{RS}Y_R = (164.13 + j10.53) \times 0.1$$

$$= 16.41 + j1.053 \text{ A} = 16.45\underline{/3.67^\circ} \text{ A}$$

$$I_Y = V_{YS}Y_Y = -9.12 - j9.47 \text{ A} = 13.15\underline{/-133.91^\circ}$$

$$I_B = V_{BS}Y_B = -7.3 + j8.42 \text{ A} = 11.14\underline{/130.9^\circ} \text{ A}$$

(d) The power consumed is

$$P_R = I_R{}^2 R_R = 16.45^2 \times 10 = 2706 \text{ W}$$

$$P_Y = I_Y{}^2 R_Y = 13.15^2 \times 20 = 3458 \text{ W}$$

$$P_B = I_B{}^2 R_B = 11.14^2 \times 25 = 3102 \text{ W}$$

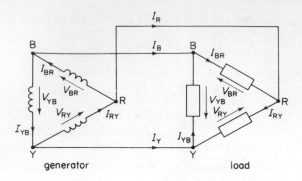

Figure 8.10   A delta-connected three-phase three-wire system

8.5 Delta Connection or Mesh Connection of Three-phase Windings

If the three windings of the generator in figure 8.2b are connected so that the 'end' of one winding is connected to the 'start' of the next winding, the result is the *delta connection* or *mesh connection* in figure 8.10.

The voltage between lines R and Y is designated by the symbol V_{RY}, where

$$V_{RY} = \text{voltage of line R relative to line Y} = V_{RN} - V_{YN}$$

The magnitude of the line voltage is given the symbol V_L and, from figure 8.4 we see that

$$\left.\begin{array}{l} V_{RY} = V_L\underline{/30^\circ} \\[2mm] V_{BR} = V_L\underline{/150^\circ} \\[2mm] V_{YB} = V_L\underline{/-90^\circ} \end{array}\right\} \qquad (8.12)$$

That it is possible to connect the winding without causing a current to circulate around the generator windings in the absence of an external load is shown in the following. Suppose that the delta connection is opened at point X in figure 8.11.

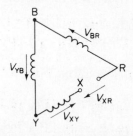

Figure 8.11   Closing the mesh connection

The voltage V_{XR} between points X and R is given by the expression

$$V_{XR} = V_{BR} + V_{YB} + V_{XY}$$

where V_{XY} has a value equal to V_{RY} in figure 8.10. Hence

$$V_{XR} = V_L \underline{/150^\circ} + V_L \underline{/-90^\circ} + V_L \underline{/30^\circ}$$

$$= V_L [(-0.866 + j0.5) - j1 + (0.866 + j0.5)]$$

$$= 0 \text{ V}$$

Since the voltage between points X and R is zero, no current flows around the loop when points X and R are connected together.

8.6 Relationship between the Line and Phase Voltages and Currents in a Delta-connected System with a Balanced Load

From figure 8.10 we see that the voltage induced in each winding (the phase voltage) is equal to the line voltage, so that in the delta system

magnitude of the phase voltage = magnitude of the line voltage

or

$$V_P = V_L \tag{8.13}$$

Applying Kirchhoff's first law to node R of the generator in figure 8.10 yields

$$I_{RY} = I_{BR} + I_R$$

from which the *line current*, I_R, is

$$\left. \begin{array}{l} I_R = I_{RY} - I_{BR} \\ \text{also at node Y } I_Y = I_{YB} - I_{RY} \\ \text{and at node B } I_B = I_{BR} - I_{YB} \end{array} \right\} \tag{8.14}$$

If the load has unity power factor, then I_{RY} is in phase with V_{RY}, I_{BR} is in phase with V_{BR}, and I_{YB} is in phase with V_{YB} to give the phasor diagram in figure 8.12. With a balanced load the magnitudes of the phase currents are equal to one another and are represented by I_P and, from figure 8.12 we see that

$$I_R = 2 I_P \cos 30^\circ = \sqrt{3} \, I_P = 1.732 \, I_P$$

Again, with a balanced load the magnitudes of I_R, I_Y and I_B are equal to one another and are equal to the line current I_L, hence

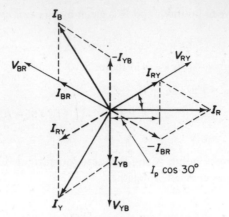

Figure 8.12   Relationship between the values of the line and phase currents in a
balanced mesh-connected system

Magnitude of the line current = $\sqrt{3}$ × Magnitude of the phase current

or $I_L = \sqrt{3}\, I_P$ (8.15)

8.7 Currents in an Unbalanced Delta-connected Load

In an unbalanced load either the magnitudes of the phase currents differ from one
another, or the phase angles differ. Equations 8.13 and 8.14 hold good for
unbalanced systems, and the general method of analysis is outlined in the following
example.

Example 8.3

A mesh-connected load consists of the following

> Impedance between lines B and R: $(10.4 + j\,6)\ \Omega$
> Impedance between lines Y and B: $(15 + j\,0)\ \Omega$
> Impedance between lines R and Y: $(7.07 - j\,7.07)\ \Omega$

If the line voltage is 300 V, calculate the values of the line currents and draw the
phasor diagram.

Solution

The phasor diagram in figure 8.13 corresponds to the solution below.

$$Z_{BR} = 10.4 + j6 = 12\underline{/30^\circ}\ \Omega$$

$$Z_{YB} = 15 + j0 = 15\underline{/0^\circ}\ \Omega$$

$$Z_{RY} = 7.07 - j7.07 = 10\underline{/-45^\circ}\ \Omega$$

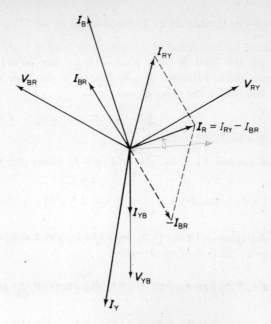

Figure 8.13

Now

$$I_{BR} = V_{BR}/Z_{BR} = 300\underline{/150^\circ}/12\underline{/30^\circ} = 25\underline{/120^\circ}\ A$$

$$= -12.5 + j21.65\ A$$

$$I_{YB} = V_{YB}/Z_{YB} = 300\underline{/-90^\circ}/15\underline{/0^\circ} = 20\underline{/-90^\circ}\ A$$

$$= -j20\ A$$

$$I_{RY} = V_{RY}/Z_{RY} = 300\underline{/30^\circ}/10\underline{/-45^\circ} = 30\underline{/75^\circ}\ A$$

$$= 7.76 + j29\ A$$

From equation 8.14

$$I_R = I_{RY} - I_{BR} = (7.76 + j29) - (-12.5 + j21.65)$$

$$= 20.26 + j7.35 = 21.55\underline{/19.94^\circ}\ A$$

$$I_Y = I_{YB} - I_{RY} = -j20 - (7.76 + j29)$$

$$= -7.76 - j49 = 49.61\underline{/-99^\circ}\ A$$

$$I_B = I_{BR} - I_{YB} = (-12.5 + j21.65) - (-j20)$$

$$= -12.5 + j41.65 = 43.48\underline{/106.71^\circ}\ A$$

8.8 Power, VA and VAr Consumed by Three-phase Systems

The power consumed by one phase of a three-phase load is $V_P I_P \cos \phi$, where ϕ is the phase angle of the load in that phase. Hence the power consumed in a three-phase system, either balanced or unbalanced, is the sum of the individual values of power consumed in the three phases.

Power consumed by a balanced system

In a *star-connected* system, $V_P = V_L/\sqrt{3}$, and $I_P = I_L$ hence the total power consumed is

$$P = 3 \times V_P I_P \cos \phi = 3 \times \frac{V_L}{\sqrt{3}} \times I_L \cos \phi = \sqrt{3} V_L I_L \cos \phi$$

In a *delta-connected* system, $V_P = V_L$, $I_P = I_L/\sqrt{3}$, and the total power consumed is

$$P = 3 \times V_P I_P \cos \phi = 3 \times V_L \times \frac{I_L}{\sqrt{3}} \times \cos \phi = \sqrt{3} V_L I_L \cos \phi$$

In general, the power consumed by a balanced load is

$$P = \sqrt{3}\ V_L I_L \cos \phi \quad \text{W}$$

Volt amperes consumed by a balanced load

The total apparent power consumed in any balanced three-phase system is

$$S = 3\ V_P I_P$$

But, from the above, $V_P I_P = V_L I_L/\sqrt{3}$, hence

$$S = \sqrt{3}\ V_L I_L \quad \text{VA}$$

Reactive volt amperes consumed by a balanced load

The total VAr consumed by a balanced load is

$$Q = 3\ V_P I_P \sin \phi = \sqrt{3}\ V_L I_L \sin \phi$$

Example 8.4

A three-phase system supplies a balanced load of 25 kW at a power factor of 0.8 lagging, the line voltage being 500 V r.m.s. Calculate the values of the line and phase currents if the load is (a) star connected, (b) delta connected.

Solution

$$P = \sqrt{3} \, V_L I_L \cos \phi$$

hence

$$I_L = P/(\sqrt{3} \, V_L \cos \phi) = 25\ 000/(\sqrt{3} \times 500 \times 0.8)$$
$$= 36 \text{ A}$$

(a) In a star-connected system

$$I_P = I_L = 36 \text{ A}$$

(b) In a delta-connected system

$$I_P = I_L/\sqrt{3} = 20.8 \text{ A}$$

Example 8.5

A delta-connected induction motor provides a shaft power of 100 kW and its power factor is 0.85 lagging. The efficiency of the motor is 86 per cent. The machine is supplied by a three-phase supply with a line voltage of 400 V, r.m.s.; calculate the value of (a) the line current and (b) the current in the motor windings.

Solution

$$\text{Power supplied} = \text{shaft power/efficiency}$$

$$= 100\ 000/0.86 = 116\ 280 \text{ W}$$

$$= \sqrt{3} V_L I_L \cos \phi$$

(a) The value of the line current is

$$I_L = 116\ 280/(\sqrt{3} \times 400 \times 0.85) = 197.4 \text{ A}$$

(b) The current in each of the delta-connected windings is

$$I_P = I_L/\sqrt{3} = 113.9 \text{A}$$

8.9 Measurement of Power in Three-phase Systems

The power consumed by an electrical system is measured by means of wattmeters, and one of the most useful types for three-phase power measurement is the *dynamometer wattmeter*. This type of instrument contains two sets of coils, one set being physically fixed to the body of the instrument and carrying the load current, *I*. Another coil which is free to rotate within the first set of coils is energised by a voltage, *V*. The pointer of the instrument indicates the value of the product *VI* cos ϕ, where ϕ is the angle between *V* and *I*. In certain instances the reading *VI* cos ϕ is not the power consumed by a specific part of the circuit. An example of this is found in the two-wattmeter method of measuring power, which is dealt with later in this section, in which the total power consumed by a three-phase load is the sum of the readings of the two wattmeters, neither wattmeter reading having significance when taken alone.

Measurement of power in a balanced three-phase system

In the special case of a balanced three-phase load, it is possible to use one wattmeter to measure the total power consumed. Two methods generally in use are shown in figure 8.14. In figure 8.14a, wattmeter W_1 reads $V_P I_L \cos \phi$, that is, the

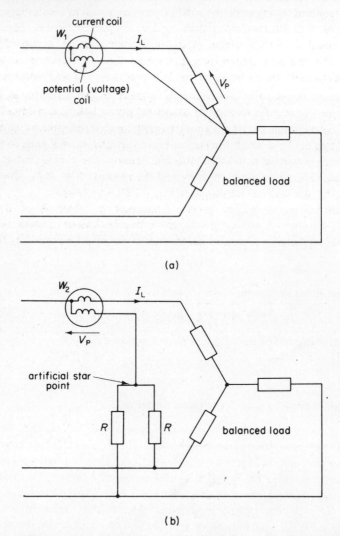

(a)

(b)

Figure 8.14 Measurement of power in a balanced three-phase load using only one wattmeter

power consumed by one phase of the load. The total power consumed is three times the reading of W_1.

Alternatively, the wattmeter is used in conjunction with two resistors which are connected so as to provide an 'artificial' neutral point. The power read by W_2 is $V_P I_L \cos \phi$, which is one-third of the power consumed by the load.

Measurement of power in any three-phase system, either balanced or unbalanced

A theorem, known as *Blondel's theorem*, states that the minimum number of wattmeters required to measure the total power consumed in a polyphase system is $(N - 1)$, where N is the number of lines used to supply the system. Thus the total power consumed in a three-phase, four-wire system can be measured by three wattmeters, and two are required for a three-phase three-wire system. An exception occurs in the special case of balanced loads, when only one wattmeter is required.

Since the most popular power-supply system is the three-phase three-wire system, the two-wattmeter method of measuring power will be considered in detail.

Typical connections for the two-wattmeter method of measuring power are shown in figure 8.15, in which the wattmeters have their current coils in two of the lines, and their potential coils are connected between the corresponding lines and the third line. If the reading of W_1 is P_1 and the reading of W_2 is P_2, then the total power taken by the load can be shown to be $P_1 + P_2$ (see below).

The instantaneous value of power consumed in phase R of the load is $p_R = v_{RS} i_R$, where v_{RS} and i_R are the instantaneous values of phase voltage and line current, respectively. The total instantaneous power p consumed by the system

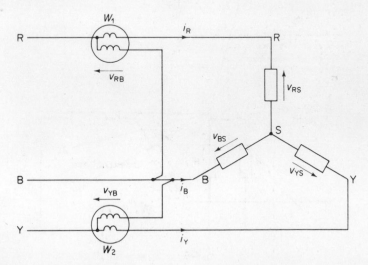

Figure 8.15 The two-wattmeter method of measuring power in a three-phase three-wire system

is

$$p = p_R + p_Y + p_B = v_{RS}i_R + v_{YS}i_Y + v_{BS}i_B$$

and the instantaneous values of power read by W_1 and W_2, respectively, are

$$p_1 = i_R v_{RB} = i_R(v_{RS} - v_{BS})$$
$$p_2 = i_Y v_{YB} = i_Y(v_{YS} - v_{BS})$$

The sum of the two wattmeter readings is

$$p_1 + p_2 = i_R v_{RS} + i_Y v_{YS} - v_{BS}(i_R + i_Y)$$

but, in a three-wire system, $i_R + i_Y = -i_B$, hence

$$p_1 + p_2 = i_R v_{RS} + i_Y v_{YS} + i_B v_{BS}$$
$$= \text{total instantaneous power} = p$$

Measurement of power in a balanced system using two wattmeters

Figure 8.16 shows the phasor diagram of the two-wattmeter circuit in figure 8.15 given that the load is balanced with a lagging phase angle. Since the load is balanced, the magnitudes of I_R and I_Y are equal to the line current I_L, and the phase angles are equal to ϕ. Also, the magnitudes of the voltages across the potential coils are both equal to V_L. The reading of W_1 in figure 8.15 is

$$P_1 = |V_{RB}I_R| \times \cos (\text{angle between } I_R \text{ and } V_{RB})$$
$$= V_L I_L \cos (30° - \phi) = V_L I_L (0.866 \cos \phi + 0.5 \sin \phi) \qquad (8.16)$$

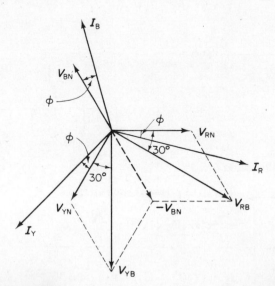

Figure 8.16   Phasor diagram for the two-wattmeter method of measuring power when the load has a lagging power factor

Note If the load has a leading power factor, then equation 8.16 becomes $P_1 = V_L I_L \cos(30° + \phi)$.

The reading of W_2 is

$$P_2 = |V_{YB} I_Y| \times \cos(\text{angle between } I_Y \text{ and } V_{YB})$$

$$= V_L I_L \cos(30° + \phi) = V_L I_L (0.866 \cos\phi - 0.5 \sin\phi) \qquad (8.17)$$

Note If the load has a leading power factor, equation 8.17 becomes $P_2 = V_L I_L \cos(30° - \phi)$. Adding equations 8.16 and 8.17 gives

$$P_1 + P_2 = \sqrt{3} V_L I_L \cos\phi$$

$$= \text{total power consumed by a balanced load} \qquad (8.18)$$

The value of angle ϕ can be deduced from the readings of W_1 and W_2 as follows. Subtracting equation 8.17 from equation 8.16 gives

$$P_1 - P_2 = V_L I_L \sin\phi$$

$$= (\text{reactive VA consumed by a balanced load})/\sqrt{3} \qquad (8.19)$$

Hence

$$\text{Reactive VA consumed by a balanced load} = Q = \sqrt{3}(P_1 - P_2)$$

Now

$$\tan\phi = \frac{\sin\phi}{\cos\phi} = \frac{(P_1 - P_2)/V_L I_L}{(P_1 + P_2)/\sqrt{3} V_L I_L} = \sqrt{3}\left(\frac{P_1 - P_2}{P_1 + P_2}\right) \qquad (8.20)$$

Also

$$\cos\phi = \frac{1}{\sec\phi} = \frac{1}{\sqrt{(1 + \tan^2\phi)}}$$

$$= \frac{1}{\sqrt{\left[1 + 3\left(\dfrac{P_1 - P_2}{P_1 + P_2}\right)^2\right]}} \qquad (8.21)$$

Special cases of balanced loads

(i) $\phi = 60°$ lagging: The total power consumed is

$$\sqrt{3} V_L I_L \cos 60° = \frac{\sqrt{3}}{2} V_L I_L$$

The reading of W_1 is

$$P_1 = V_L I_L \cos(30° - 60°) = V_L I_L \cos 30° = \frac{\sqrt{3}}{2} V_L I_L$$

$$= \text{total power consumed by the load}$$

and W_2 reads

$$P_2 = V_L I_L \cos(30° + 60°) = 0$$

(ii) $\phi > 60°$ lagging: The reading of W_2 is negative and W_1 gives a positive reading which is greater than the power consumed by the load.

(iii) $\phi = 60°$ leading: The total power consumed is

$$\sqrt{3} V_L I_L \cos 60° = \frac{\sqrt{3}}{2} V_L I_L$$

The reading of W_1 is

$$P_1 = V_L I_L \cos[30° - (-60°)] = 0$$

and the reading of W_2 is

$$P_2 = V_L I_L \cos[30° + (-60°)] = \frac{\sqrt{3}}{2} V_L I_L$$

$$= \text{total power consumed by the load}$$

(iv) $\phi > 60°$ leading: The reading of W_1 is negative, and W_2 gives a positive reading which is greater than the power consumed by the load.

Example 8.6

A balanced load of 12 kVA is connected to a three-phase three-wire system. Determine the readings of two wattmeters connected as shown in figure 8.15 if the power factor of the load is (a) unity, (b) 0.866 lagging, and (c) 0.5 leading. What is the maximum possible reading of either wattmeter, and what is the reading of each instrument for zero lagging power factor?

Solution

The total power consumed is

$$\text{kVA} \times \text{power factor} = 12 \cos \phi \text{ kW} = P_1 + P_2 \qquad (8.22)$$

Substituting equation 8.22 into equation 8.20 gives

$$\tan \phi = \sqrt{3}(P_1 - P_2)/(12 \cos \phi)$$

or

$$P_1 - P_2 = 12 \cos \phi \tan \phi/\sqrt{3} = 12 \sin \phi/\sqrt{3}$$

(a) $\cos \theta = 1.0 \ (\phi = 0°)$

$$P_1 - P_2 = 12 \times 1 \times 0/\sqrt{3} = 0$$

or $P_1 = P_2$, hence $P_1 = 6$ kW and $P_2 = 6$ kW.

(b) $\cos \phi = 0.866$ (lagging) ($\phi = 30°$ lagging)

$$P_1 - P_2 = 12 \sin 30°/\sqrt{3} = 12 \times 0.5/\sqrt{3} = 3.46 \qquad (8.23)$$

and

$$P_1 + P_2 = 12 \cos \phi = 12 \times 0.866 = 10.392 \qquad (8.24)$$

Solving between equations 8.23 and 8.24 for P_1 and P_2 yields

$$P_1 = 6.93 \text{ kW and } P_2 = 3.462 \text{ kW}$$

(c) $\cos \phi = 0.5$ (leading) ($\phi = 60°$ leading)

Note Since ϕ leads, then $\sin \phi$ has a negative value.

$$P_1 - P_2 = 12 \sin (-60°)/\sqrt{3} = 12 \times (-0.866)/\sqrt{3}$$
$$= -6 \text{ kW} \qquad (8.25)$$

and

$$P_1 + P_2 = 12 \cos \phi = 12 \times 0.5 = 6 \text{ kW} \qquad (8.26)$$

Solving between equations 8.25 and 8.26 yields

$$P_1 = 0 \text{ kW and } P_2 = 6 \text{ kW}$$

The maximum reading on either wattmeter occurs when the current flowing in the current winding of the wattmeter is in phase with the voltage across the voltage winding. This occurs for W_1 at a phase angle of $30°$ lagging, and for W_2 at a phase angle of $30°$ leading. In either case the maximum reading is 6.93 kW (see solution b above).

When the power factor is zero ($\phi = 90°$)

$$P_1 + P_2 = 12 \cos 90° = 0 \qquad (8.27)$$

and

$$P_1 - P_2 = \frac{12}{\sqrt{3}} \sin 90° = 6.93 \qquad (8.28)$$

Solving between equations 8.27 and 8.28 yields

$$P_1 = 3.465 \text{ kW} \quad \text{and} \quad P_2 = -3.465 \text{ kW}$$

8.10 Measurement of Reactive VA in Three-phase Three-wire Systems

From equation 8.19

$$V_L I_L \sin \phi = P_1 - P_2$$

hence the reactive VA consumed by the system is

$$Q = \sqrt{3} \, V_L I_L \sin \phi = \sqrt{3}(P_1 - P_2)$$

In the special case of a balanced load, a single wattmeter may be used to measure the reactive VA, provided that the coil is connected in series with one line (say the R line) and that the potential coil is connected between the other pair of lines (the Y and B lines). The reading of the wattmeter is then $V_L I_L \sin \phi$, which value must be multiplied by a factor of $\sqrt{3}$ to give the total reactive VA consumed.

8.11 Power Factor Correction in Three-phase Systems

As stated in chapter 6, for a given power consumption a system with a low power factor draws a larger current than one with a higher power factor. Industrial consumers usually have a large number of induction motor drives whose starting power factor may be about 0.3 lagging, and whose light load power factor may be 0.6 lagging. One method adopted to improve the power factor of lagging loads is to connect a three-phase capacitor bank to the terminals of the load, the VAr rating of the capacitor being chosen to give an optimum overall power factor. In some cases it is possible to improve the power factor of the system as a whole by installing motor drives which draw a leading current from the power supply; synchronous motors and synchronous induction motors are machines of this type.

Example 8.7

A star-connected three-phase 440-V, 50-Hz induction motor takes a line current of 40 A at 0.8 power factor lagging. A three-phase delta-connected capacitor bank is used to raise the overall power factor to 0.95 lagging. Calculate the kVA rating of the capacitor bank, and also determine the value of the capacitance used in each phase of the capacitor bank.

Solution

A phasor diagram showing the line currents involved is given in figure 8.17. In this diagram, I_{L1} is the line current drawn by the induction motor, I_{L2} is the component of line current due to the capacitor bank, and I_L is the total line current at a power factor of 0.95 lagging. From the values given

$$\phi_1 = \cos^{-1} 0.8 = 36.87° \quad \text{and} \quad \phi = \cos^{-1} 0.95 = 18.19°$$

The quadrature component I_{Q1} of line current I_{L1} is

$$I_{Q1} = I_{L1} \sin \phi = 40 \times 0.6 = 24 \text{ A}$$

and the 'in-phase' component of I_{L1} is

$$I_{L1} \cos \phi_1 = 40 \times 0.8 = 32 \text{ A}$$

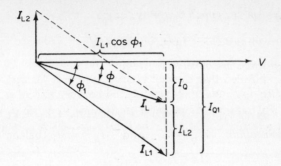

Figure 8.17

The quadrature component I_Q of the total line current I_L is

$$I_Q = (I_{L1} \cos \phi) \tan \phi = 32 \times 0.3288 = 10.52 \text{ A}$$

Hence the component of current I_{L2} flowing into the capacitor is

$$I_{L2} = I_{Q1} - I_Q = 24 - 10.52 = 13.48 \text{ A}$$

The kVAr rating of the capacitor bank is

$$Q = \sqrt{3} \, V_L I_{L2}/1000 = \sqrt{3} \times 440 \times 13.48/1000$$
$$= 10.27 \text{ kVAr}$$

If I_P is the current flowing into each phase of the delta-connected capacitor bank, then

$$I_P = I_{L2}/\sqrt{3} = 13.48/\sqrt{3} = 7.78 \text{ A}$$

Now, $I_P = V_L \omega C$, where C is the capacitance of one phase of the capacitor; hence

$$C = I_P/V_L \omega = 7.78/440 \times 2\pi \times 50 \text{ F} = 56.27 \ \mu\text{F}$$

Summary of essential formulae

Three-phase voltages: $\quad V_{RN} = V_P\underline{/0^\circ}$

$$V_{YN} = V_P\underline{/-120^\circ} = V_P(-0.5 - j0.866)$$

$$V_{BN} = V_P\underline{/120^\circ} = V_P(-0.5 + j0.866)$$

$$V_{RY} = V_{RN} - V_{YN}$$

$$V_{YB} = V_{YN} - V_{BN}$$

$$V_{BR} = V_{BN} - V_{RN}$$

Balanced three-phase supply: $\quad V_L = \sqrt{3} V_P$

Star connection: $V_L = \sqrt{3}V_P$ and $I_L = I_P$

Three-phase four-wire system: $I_R + I_Y + I_B = I_N$

Three-phase three-wire star-connected system:

$$I_R + I_Y + I_B = 0$$

$$V_{SN} = (V_R Y_R + V_Y Y_Y + V_B Y_B)/(Y_R + Y_Y + Y_B)$$

Delta or mesh connection: $V_L = V_P$ and $I_L = \sqrt{3}I_P$ (balanced load)

$$I_R = I_{RY} - I_{BR}$$

$$I_Y = I_{YB} - I_{RY}$$

$$I_B = I_{BR} - I_{YB}$$

Three-phase balanced load:

volt amperes: $S = 3V_P I_P = \sqrt{3}V_L I_L$ VA

reactive volt amperes: $Q = 3V_P I_P \sin\phi = \sqrt{3}V_L I_L \sin\phi$ VAr

power: $P = 3V_P I_P \cos\phi = \sqrt{3}V_L I_L \cos\phi$ W

Two-wattmeter method:

total power $= P_1 + P_2$

reactive VA $= \sqrt{3}(P_1 - P_2)$

phase angle $= \phi = \tan^{-1}[\sqrt{3}(P_1 - P_2)/(P_1 + P_2)]$

PROBLEMS

8.1 A three-phase alternator supplies a balanced three-phase load with a line current of 50 A at a line voltage of 550 V. Determine (a) the line-to-neutral voltage and (b) the apparent power supplied by the alternator.
[(a) 317.5 V; (b) 47.63 kVA]

8.2 A three-phase balanced load absorbs 50 kW at a power factor of (a) 0.6 lagging, (b) 0.8 leading. If the line voltage is 500 V, calculate the line current in each case.
[(a) 96.22 A; (b) 72.17 A]

8.3 A three-phase, star-connected alternator generates a line voltage of 11 kV. The alternator supplies a balanced load of 20 MVA at a power factor of 0.75 lagging. Determine (a) the line-to-neutral voltage, (b) the power output, (c) the line current and (d) the reactive kVA supplied to the load.
[(a) 6.35 kV; (b) 15 MW; (c) 1050 A; (d) 13.23 kVAr]

8.4 A three-phase, four-wire, 440-V system supplies a star-connected load in which the following are connected

R–N   resistance of 10 Ω

Y–N   a coil of impedance 15 Ω and phase angle 30°

B–N C and R in series of impedance 10 Ω and phase angle $-30°$

Determine the magnitude and phase angle of each line current and of the neutral wire current. Calculate also the power consumed by the load.

[I_R = 25.4 A in phase with V_{RN}; I_Y = 16.93 A lagging V_{YN} by 30°; I_B = 25.4 A leading V_{BN} by 30°; I_N = 12.03 A leading V_{RN} by 159.4°; 15.763 kW]

8.5 A symmetrical three-phase generator supplies an unbalanced four-wire, star-connected load with the following phase impedances: Z_{RN} = 100 + j0 Ω; Z_{YN} = 100 + j30 Ω; Z_{BN} = 75 −j130 Ω. Determine the current in each line and also in the neutral wire given that V_{RN} = 100 + j0 V. Calculate the total power consumption.

If Z_{RN} and Z_{BN} remain unchanged, determine an expression for the complex impedance Z_{YN} which causes the neutral wire current to be zero.

[I_R = 1 $\underline{/0°}$ A; I_Y = 0.96 $\underline{/-136.7°}$ A; I_B = 0.6667 $\underline{/180°}$ A; I_N = 0.751 $\underline{/-119°}$A; 225 W; 300 $\underline{/60°}$ Ω]

8.6 Two wattmeters are used to measure the power consumed by a three-phase, three-wire system. The current coil of one of the wattmeters (P_1) is connected in series with the R-line, and the current coil of the second wattmeter (P_2) is connected in series with the Y-line. Draw the wiring diagram of the arrangement and also draw the phasor diagram if the load has a lagging power factor.

A 420-V, three-phase, three-wire, star-connected load draws an R-line current of 10 $\underline{/-30°}$ A (relative to V_{RN}) and a Y-line current of 14 $\underline{/-60°}$ A (relative to V_{YN}). Determine the current in line B and the reading of wattme'ers P_1 and P_2.

[I_B = 23.2 A leading V_{BN} by 12.45°; P_1 = 4200 W; P_2 = 0]

8.7 A delta-connected load is supplied from a 400-V, three-phase, three-wire system of phase sequence RYB. The load impedances are: Z_{RY} = 60 + j80 Ω; Z_{YB} = 50 + j0 Ω; Z_{BR} = 50 − j50 Ω. Determine the magnitude and phase angle of each line current, and calculate the power consumed by each phase of the load and also the total power consumed.

[I_R = 9.15 $\underline{/-0.69°}$ A; I_Y = 7.4 $\underline{/-120°}$ A; I_B = 8.53 $\underline{/130°}$ A; P_{BR} = 1602 W; P_{RY} = 960 W; P_{YB} = 3200 W; P_T = 5762 W]

8.8 Two wattmeters P_1 and P_2 are used to measure the power consumed by a balanced three-phase load. Determine the power factor of the load when (a) P_1 = 2000 W, P_2 = 500 W, (b) P_1 = 2000 W, P_2 = −500 W.

[(a) 0.6934; (b) 0.3273]

8.9   The following electrical loads are connected to a supply system

>   21 kW at a power factor of 0.707 lagging
>   32 kW at a power factor of 0.8 lagging
>   18.8 kW at a power factor of 0.94 leading
>   36 kW at a power factor of 0.9 lagging

Determine the total value of (a) the kVA and (b) the reactive kVA consumed. Calculate also the reactive kVA required to improve the overall power factor to 0.97 lagging.

If the line voltage of the three-phase supply is 550 V at a frequency of 50 Hz, and the power factor improvement is brought about by a delta-connected bank of capacitors, determine the capacitance in μF of each phase of the capacitor bank.

[(a) 121.3 kVA; (b) 55.61 kVAr; 28.6 kVAr; 100.2 μF]

9 *Transformers and Coupled Circuits*

9.1 Introduction

An important feature of a.c. distribution systems is the ease with which the system voltage can be either increased or reduced by means of transformers. When transmitting a given amount of power, a high supply voltage results in a small line current with consequent small *copper losses* (I^2R losses). Thus, when large amounts of power are transmitted over great distances, it is done at a very high voltage, typically several hundred kilovolts. In local distribution networks the line voltage may be 11 or 6.6 kV, and in medium and large sizes of industrial installations power is distributed at 3.3 kV. The voltage at which the supply is utilised depends on the nature of the installation, and may be at a line voltage of 3.3 kV in the case of large machines or 415 V in smaller machines. Intermediate voltages such as 1.1 kV are used in some industries.

A popular distribution voltage for domestic use in the United Kingdom is 415 V three-phase, which is finally connected to consumer's terminals in the form of a 240-V single-phase supply. Portable tools in industry are frequently supplied at 110 V, the transformer secondary having its centre-point earthed to give a voltage of 55 V between either line and earth.

Transformers are both costly and bulky, and the electronics industry endeavours to utilise circuitry which avoids their use. Nevertheless, their use is unavoidable either where circuits require a non-standard supply voltage or where the circuit must be electrically isolated from the power supply.

Transformers are also used as a means of changing the apparent impedance of a load (see section 9.3), so that the impedance of a load as 'seen' by the power supply is either larger or smaller than its actual value. As we saw with the maximum power transfer theorem (section 2.6), the load consumes its maximum power when the impedance of the load is equal to the impedance of the supply source. By using a transformer as an 'impedance matching' device, it is possible to modify the apparent impedance of the load, which may be a loudspeaker, so that maximum power may be transferred into it.

9.2 Principle of Operation

A simplified sketch of a transformer is shown in figure 9.1; it consists of two separate windings mounted on a *laminated iron core.* The purpose of laminating the core is to reduce the magnitude of the *eddy currents* induced in the iron by the magnetic flux alternations. Individual laminations are insulated from one another by a layer of varnish, paper or other insulant. The winding that is energised by the power supply is known as the *primary winding*, and the winding to which the load is connected is the *secondary winding.* In the figure, both windings are on the vertical parts of the core, known as the *limbs* of the transformer. The horizontal parts of the core completing the magnetic circuit between the limbs are known as the *yokes.* In small transformers the core is rectangular in section, and the laminations are held in place by the windings or, in some cases, by tape. In medium and large transformers the sizes of the laminations are stepped to give an approximately circular section, and are clamped together by bolts which pass through the core but are insulated from it.

The dot notation (see also chapter 3) applied to figure 9.1 indicates that when point A on the primary winding is positive with respect to point B then, in the secondary, point D is positive with respect to point C. Since the two windings are electrically isolated from one another, it is not possible to deduce the phase relationship between V_1 and V_2 until a common reference node has been selected. If terminals B and C are linked by the dotted connection shown in the figure, we see that when terminal A is positive with respect to the common link, terminal D is also positive with respect to the common connection. Had the common connection

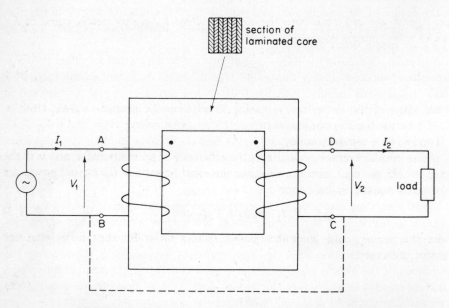

Figure 9.1   Basic features of a single-phase transformer

been made between B and D (or between A and C), then the primary and secondary voltages would have been antiphase to one another. Thus the phase relationship between the voltages induced in the windings depends not only on the directions of the windings but also on the choice of the reference node.

It was shown in chapter 3 (see equation 3.2) that the instantaneous value of the induced e.m.f., e, in a winding of N turns is

$$e = N \, d\Phi/dt$$

where $d\Phi/dt$ is the rate of change of the flux linking with the coil. Hence

$$\frac{e}{N} = \frac{d\Phi}{dt}$$

Under no-load conditions the magnetic flux links both windings, and if the flux waveform is sinusoidal, then the ratio

r.m.s. value of induced e.m.f. : number of turns

is the same in both windings. If E_1 is the r.m.s. voltage induced in the primary winding and E_2 is the r.m.s. voltage induced in the secondary winding, then

$$\frac{E_1}{N_1} = \frac{E_2}{N_2}$$

or

$$\frac{E_2}{E_1} = \frac{N_2}{N_1} \tag{9.1}$$

Also, under no-load conditions the applied voltage V_1 is very nearly equal to E_1, and V_2 is very nearly equal to E_2 so that

$$\frac{V_2}{V_1} \approx \frac{N_2}{N_1} \tag{9.2}$$

If the value of the secondary voltage is lower than the primary voltage (that is, $V_2 < V_1$), the transformer is said to have a *step-down* voltage ratio, and if $V_2 > V_1$ it is said to have a *step-up* ratio.

Under normal operating conditions the efficiency of power transformers is in the range 95–98 per cent, and the values of the input power and the output power are very nearly equal in value, hence

$$V_1 I_1 \cos \phi_1 = V_2 I_2 \cos \phi_2 \tag{9.3}$$

Since the primary and secondary power factors differ by very little from one another, then

$$\frac{V_2}{V_1} \approx \frac{I_1}{I_2} \tag{9.4}$$

Combining equations 9.2 and 9.4 yields

$$\frac{V_2}{V_1} \approx \frac{N_2}{N_1} \approx \frac{I_1}{I_2} \qquad (9.5)$$

From equation 9.5 we see that

$$I_1 N_1 = I_2 N_2 \qquad (9.6)$$

That is, *the number of ampere turns on the secondary winding resulting from the load current is balanced by an equal number of ampere turns in the primary winding*. Thus an increase in load current produces a corresponding increase in primary current.

Example 9.1

A single-phase step-down transformer of ratio 5:1 delivers a secondary current of 55 A at a power factor of 0.9 lagging. Calculate the magnitude of the component of primary current resulting from the flow of load current.

Solution

From the data given, $I_2 = 55$ A and $N_2/N_1 = 1/5 = 0.2$. From equation 9.6

$\qquad I_1 = I_2 N_2/N_1 = 55 \times 0.2 = 11$ A at a lagging power factor of 0.9

9.3 The Transformer as an Impedance-level Converting Device

In electronic circuits, transformers are sometimes used as impedance-level converting devices. Assuming that the transformer in figure 9.2 is an ideal loss-free transformer, then

$$I_2 = V_2/R_L$$

or

$$R_L = V_2/I_2 \qquad (9.7)$$

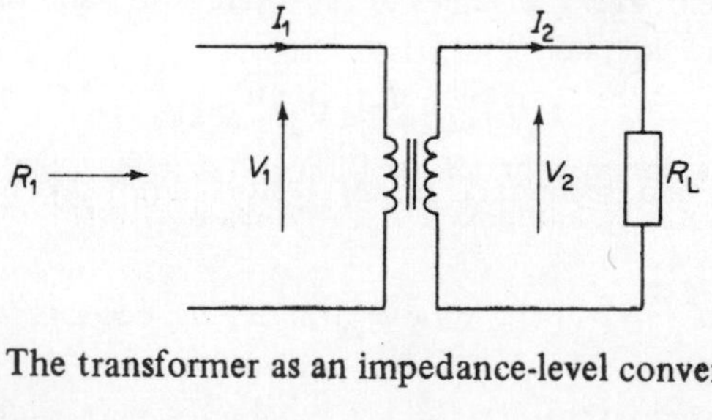

Figure 9.2   The transformer as an impedance-level converting device

For an ideal transformer

$$I_1 = \frac{N_2}{N_1} I_2$$

and

$$V_1 = \frac{N_1}{N_2} V_2$$

The apparent resistance, R_1, seen by the primary winding supply source is

$$R_1 = \frac{V_1}{I_1} = \left(\frac{N_1}{N_2} V_2\right) \times \left(\frac{N_1}{N_2 I_2}\right) = \left(\frac{N_1}{N_2}\right)^2 \frac{V_2}{I_2}$$

Substituting equation 9.7 into the above expression yields

$$R_1 = \left(\frac{N_1}{N_2}\right)^2 \times R_L \qquad (9.8)$$

That is, the resistance between the primary winding terminals is $(N_1/N_2)^2$ times the value of the load resistance.

Example 9.2

A resistance of $15\,\Omega$ is connected to the secondary of a transformer with a step-down voltage ratio of $5:1$. Compute the effective resistance to flow of primary current.

Solution

$N_1/N_2 = 5$, therefore from equation 9.8

$$R_1 = 5^2 \times 15 = 375\,\Omega$$

9.4 E.M.F. Equation of the Transformer

Since the magnitude of the magnetic flux in the core of the transformer varies sinusoidally, its equation is

$$\Phi = \Phi_m \sin \omega t$$

where Φ_m is the maximum value of the flux, and ω is the angular frequency of the supply in rad/s. The e.m.f., e, induced in a winding of N turns on the core is

$$e = N \frac{d\Phi}{dt} = Nd(\Phi_m \sin \omega t)/dt = \omega N \Phi_m \cos \omega t \quad \text{volts}$$

$$= \omega N \Phi_m \sin (\omega t + 90°) = E_m \sin (\omega t + 90°) \text{ volts} \qquad (9.9)$$

That is to say, the maximum value of induced e.m.f., E_m, is

$$E_m = \omega N \Phi_m \qquad (9.10)$$

and the e.m.f. phasor *leads* the magnetic flux phasor by 90°. Since the induced e.m.f. waveform is sinusoidal, its r.m.s. value, E, is

$$E = E_m / \sqrt{2} = \omega N \Phi_m / \sqrt{2} = 2\pi f N \Phi_m / \sqrt{2}$$

$$= 4.44 f N \Phi_m \quad \text{volts} \qquad (9.11)$$

The r.m.s. values of the e.m.f.s E_1 and E_2 induced in the primary and secondary windings, respectively, are

$$E_1 = 4.44 f N_1 \Phi_m \qquad (9.12)$$

$$E_2 = 4.44 f N_2 \Phi_m \qquad (9.13)$$

Example 9.3

The primary winding of a single-phase transformer is energised by a 230-V, 50-Hz supply. If the maximum value of core flux is 0.001802 Wb, the secondary winding has 1150 turns and the maximum flux density is 0.36 **tesla**, calculate (a) the number of turns on the primary winding, (b) the secondary induced voltage, and (c) the net cross-sectional area of the core.

Solution

From the data provided, $E_1 \approx V_1 = 230$ V, $f = 50$ Hz, $\Phi_m = 0.001802$ Wb, $N_2 = 1150$ turns, and $B_m = 0.36$ T.

(a) From equation 9.12

$$N_1 = E_1 / 4.44 f \Phi_m = 230/4.44 \times 50 \times 0.001802 = 575 \text{ turns}$$

(b) From equation 9.13

$$E_2 = 4.44 f N_2 \Phi_m = 4.44 \times 50 \times 1150 \times 0.001802 = 460 \text{ V}$$

Note: E_2 could have been calculated from $E_2 = E_1 N_2 / N_1 = 230 \times 1150/575 = 460$ V.

(c) $$B_m = \Phi_m / \text{core area}$$

therefore

$$\text{core area} = \phi_m / B_m = 0.001\,802/0.36 = 0.005 \text{ m}^2 = 50 \text{ cm}^2$$

9.5 No-load Phasor Diagram for the Transformer Neglecting the Voltage Drop in the Primary Winding

Following the usual practice when drawing phasor diagrams, the quantity common to all parts of the magnetic circuit is shown on the reference (horizontal) axis. In the transformer this quantity is the magnetic flux, as shown in figure 9.3. From equation 9.9, the induced e.m.f.s in both the primary and secondary windings lead the flux phasor by $90°$ and, assuming that the turns ratio is 1:1, then $E_1 = E_2$. Also, since the effects of the voltage drops in the windings are neglected, $V_1 = E_1$ and, since no current is drawn from the secondary, $V_2 = E_2$. In a practical transformer the turns ratio is other than 1:1, and the value of E_2 is $E_1 N_2/N_1$.

In order to maintain the magnetic flux in the core, the primary winding carries a current I_0, known as the *no-load current*, which lags behind the supply voltage by angle ϕ_0. This current is regarded as consisting of two components, namely the *magnetising component*, I_{mag}, which lags behind V_1 by $90°$, and the *core loss component*, I_C, which is in phase with V_1. The latter component gives rise to a power loss of $V_1 I_C$ in the core, and is known as the *iron loss* or *core loss*, P_0. Hence

$$\left.\begin{aligned}
I_C &= I_0 \cos \phi_0 \\
I_{mag} &= I_0 \sin \phi_0 \\
I_0 &= \sqrt{(I_C{}^2 + I_{mag}{}^2)} \\
\phi_0 &= \tan^{-1}(I_{mag}/I_C) = \cos^{-1}(I_C/I_0) \\
\text{core loss} &= P_0 = V_1 I_C = V_1 I_0 \cos \phi_0
\end{aligned}\right\} \qquad (9.14)$$

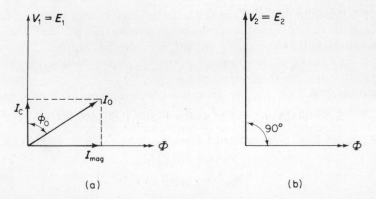

(a)   (b)

Figure 9.3   No-load phasor diagrams for a single-phase transformer: (a) primary winding, (b) secondary winding

9.6 Phasor Diagram for the Transformer under Loaded Conditions Neglecting the Voltage Drops in the Windings

Suppose that an inductive load is connected to the secondary winding. The phasor diagram for the secondary winding is shown in figure 9.4b. The secondary current, I_2, lags behind E_2 by ϕ_2, and a corresponding current I_1' flows in the primary winding to maintain ampere-turn balance, where

$$I_1' = I_2 N_2 / N_1$$

The total primary current, I_1, is the phasor sum of I_1' and I_0; see figure 9.4.

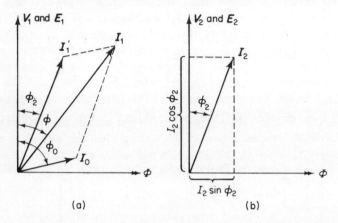

(a)   (b)

Figure 9.4   Transformer phasor diagrams for a loaded transformer (neglecting the voltage drops in the windings): (a) primary winding, (b) secondary winding

The equivalent electrical circuit which represents the phasor diagram in figure 9.4 is shown in figure 9.5, in which the primary current I_1 divides into the no-load current I_0 and a current I_1', the latter flowing into an 'ideal' transformer which requires neither magnetising current nor core loss current. The core loss current, I_C, is assumed to flow through resistor R_C, and the magnetising current is assumed to flow through inductive reactance X_m.

The values of R_C and X_m are determined from a *no-load test* or *open-circuit test*, in which the primary voltage V_1, the primary no-load current I_0, and the primary no-load power (the core loss) P_C are measured. From these data

$$\left. \begin{aligned} \cos \phi_0 &= P_0 / V_1 I_0 \\ I_C &= I_0 \cos \phi_0 \\ I_{mag} &= I_0 \sin \phi_0 \\ R_C &= V_1 / I_C \text{ and } X_m = V_1 / I_{mag} \end{aligned} \right\} \tag{9.15}$$

hence

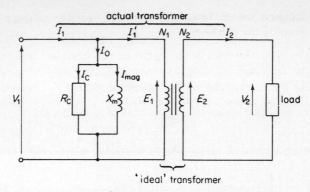

actual transformer

'ideal' transformer

Figure 9.5 One method of making allowances for the no-load current I_0 in the equivalent circuit of the transformer

Example 9.4

A single-phase transformer with a step-down voltage ratio of 10:1 draws a primary current of 5.5 A at a power factor of 0.9 lagging when the secondary current is 50 A at a power factor of 0.95 lagging. Determine the magnitude and the power factor of the no-load current. Neglect the effect of the voltage drops in the windings.

Solution

$$\phi_1 = \cos^{-1} 0.9 = 25.84°$$

$$\sin \phi_1 = 0.4358$$

$$\phi_2 = \cos^{-1} 0.95 = 18.19°$$

$$\sin \phi_2 = 0.3123$$

From figure 9.4b

$$I_2 \cos \phi_2 = 50 \times 0.95 = 47.5 \text{ A}$$

$$I_2 \sin \phi_2 = 50 \times 0.3123 = 15.62 \text{ A}$$

The perpendicular component of I_1' in the primary winding is

$$I_1' \cos \phi_2 = \left(\frac{N_2}{N_1}\right) I_2 \cos \phi_2 = 0.1 \times 47.5 = 4.75 \text{ A}$$

and the horizontal component of I_1' is

$$I_1' \sin \phi_2 = \left(\frac{N_2}{N_1}\right) I_2 \sin \phi_2 = 0.1 \times 15.62 = 1.562 \text{ A}$$

The perpendicular component of I_1 is

$$I_1 \cos \phi_1 = 5.5 \times 0.9 = 4.95 \text{ A}$$

and the horizontal component of I_1 is

$$I_1 \sin \phi_1 = 5.5 \times 0.4358 = 2.4 \text{ A}$$

Hence the core loss (perpendicular) component of I_0 is

$$I_C = I_0 \cos \phi_0 = I_1 \cos \phi_1 - I_1' \cos \phi_2 = 4.95 - 4.75$$

$$= 0.2 \text{ A}$$

and the magnetising (horizontal) component of I_0 is

$$I_{mag} = I_0 \sin \phi_0 = I_1 \sin \phi_1 - I_1' \sin \phi_2$$

$$= 2.4 - 1.562 = 0.838 \text{ A}$$

From equation 9.14

$$I_0 = \sqrt{(I_C^2 + I_{mag}^2)} = \sqrt{(0.2^2 + 0.838^2)} = 0.861 \text{ A}$$

and

$$\cos \theta_0 = I_C/I_0 = 0.2/0.861 = 0.2323$$

Example 9.5

In an open-circuit test on a single-phase transformer it was found that, when 200 V was applied to the primary winding, the no-load current was 0.7 A and the power consumed was 70 W. Calculate the values of R_C and X_m in the primary circuit (see figure 9.5).

Solution

From equation 9.15

$$\cos \phi_0 = P_0/V_1 I_0 = 70/(200 \times 0.7) = 0.5$$

and

$$\sin \phi_0 = \sqrt{(1 - 0.5^2)} = 0.866$$

The components of I_0 are

$$I_C = I_0 \cos \phi_0 = 0.7 \times 0.5 = 0.35 \text{ A}$$

and

$$I_{mag} = I_0 \sin \phi_0 = 0.7 \times 0.866 = 0.606 \text{ A}$$

Therefore

$$R_C = V_1/I_C = 200/0.35 = 571 \; \Omega$$

and

$$X_m = V_1/I_{mag} = 200/0.606 = 330 \; \Omega$$

9.7 Leakage Flux in a Transformer

When a resistive load is connected to the secondary winding of a transformer, in the manner shown in figure 9.6, a current I_2 flows in the load. The direction of flow of the secondary current can be deduced from Lenz's law (see section 3.2), which states that the current circulates in a direction which opposes the change of flux causing the secondary current. With the instantaneous direction of flow of the primary current shown in the figure, the useful magnetic flux Φ circulates in a clockwise direction around the core. It follows from Lenz's law that the secondary current circulates in a direction to produce a flux Φ_{L2} which opposes the main flux. The magnitude of this flux is proportional to the value of the load current.

The net result is that the main flux tends to be reduced, and with it the induced 'back' e.m.f. E_1 in the primary winding also tends to reduce. Consequently, the primary current increases to a value which produces ampere-turn balance on the two windings. As a result, the useful flux Φ remains substantially constant over the working load range of the transformer, and the component of current in the primary due to the load causes a *leakage flux* Φ_{L1} to be established around the primary coil. Flux Φ_{L2} associated with the flow of load current in the secondary is known as the secondary winding leakage flux.

Since flux Φ_{L1} links only with the primary winding, it does not contribute to

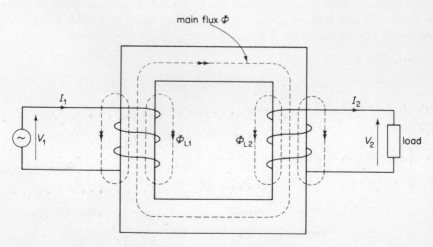

Figure 9.6   Leakage flux paths

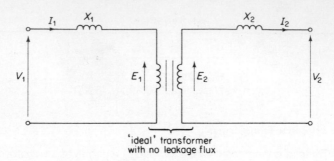

Figure 9.7   Representation of leakage reactances on the equivalent circuit of the transformer

the output voltage, and has the effect of inducing a further 'back' e.m.f. in the primary winding. Consequently the voltage drop caused by the leakage flux Φ_{L1} can be regarded as being produced by a *leakage reactance X_1* in series with the primary winding, shown in figure 9.7. Also, the voltage drop associated with the secondary leakage flux can also be regarded as being produced by a leakage reactance X_2 in series with the secondary winding.

9.8 Approximate Equivalent Circuit of a Single-phase Transformer

In addition to the voltage drops due to flow of current in X_1 and X_2, current also flows through the resistances R_1 and R_2 of the primary and secondary windings, respectively. An approximate equivalent circuit of the transformer is shown in figure 9.8, in which Z_1 is the primary winding impedance and includes the resistance and the leakage reactance of the primary winding. The secondary winding impedance is Z_2. This diagram neglects the effects of the primary no-load current (see figures 9.5 and 9.9). In power transformers the values of X_1 and X_2 are about three to eight times the values of R_1 and R_2, respectively.

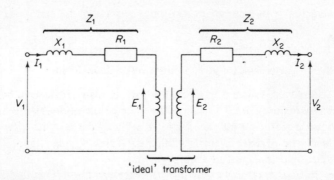

Figure 9.8   Approximate equivalent circuit of the transformer

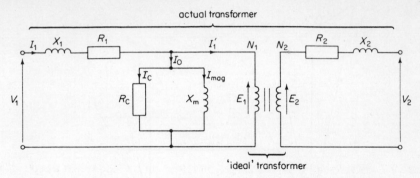

Figure 9.9 Complete equivalent circuit of the single-phase transformer

9.9 Complete Equivalent Circuit of a Single-phase Transformer

The operation of a practical power transformer can be expressed in terms of the equivalent circuit in figure 9.9, in which the winding impedances and the effects of the no-load current are accounted for by components connected external to an 'ideal' transformer, whose iron circuit has an infinite permeability and requires no magnetising current.

The phasor diagram for the complete equivalent circuit is shown in figure 9.10. The secondary terminal voltage is evaluated by solving the equation

$$V_2 = E_2 - I_2 Z_2 = E_2 - I_2 (R_2 + jX_2)$$

where $I_2 Z_2$ is the voltage drop in the secondary winding impedance. This subtraction is performed in the manner shown in the secondary circuit phasor diagram, figure 9.10b, in which $-I_2 R_2$ is antiphase to I_2, and $-I_2 X_2$ leads the phasor $(-I_2)$ by $90°$. The primary current is evaluated from the phasor sum of I_1' and I_0. The equation for the primary terminal voltage is

$$V_1 = E_1 + I_1 Z_1 = E_1 + I_1 (R_1 + jX_1)$$

The above addition is performed on the phasor diagram for the primary circuit, figure 9.10a.

9.10 Simplified Equivalent Circuit of a Single-phase Transformer

The complete equivalent circuit in figure 9.9 can, in many instances, be simplified to one of the circuits in figure 9.11. Since the value of the no-load current is usually less than about 5 per cent of the full-load primary current, its effects may be neglected. Hence the parallel circuit containing R_C and X_m (see figure 9.9) may be omitted without significant loss of accuracy.

Moreover, it is theoretically possible to refer all the resistance and all the reactance of the transformer into one or other of the windings, so that the

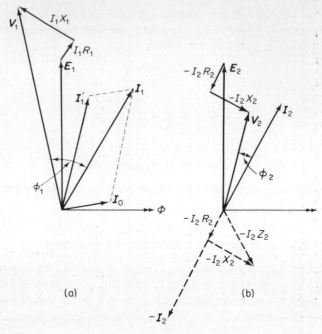

Figure 9.10 Phasor diagram for the complete equivalent circuit: (a) primary circuit, (b) secondary circuit

remaining winding has neither resistance nor reactance. In figure 9.11a, R_{E1} is the effective resistance of the whole transformer *referred to the primary winding*, where

$$R_{E1} = R_1 + R_2' \qquad (9.16)$$

That is, R_{E1} is the sum of the resistance R_1 of the primary winding and a resistance R_2', which is the *resistance of the secondary winding referred to the primary*

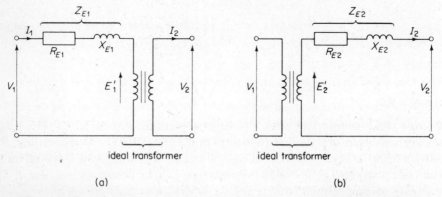

Figure 9.11 Simplified equivalent circuit of the single-phase transformer

winding. If the referred value of resistance R_2' is to produce the same effect in the primary winding as R_2 does in the secondary winding, then each must absorb the same amount of power in the circuit. That is

$$I_1{}^2 R_2' = I_2{}^2 R_2$$

or

$$R_2' = R_2 (I_2/I_1)^2 = R_2 (N_1/N_2)^2 \tag{9.17}$$

Hence

$$R_{E1} = R_1 + R_2 (N_1/N_2)^2 \tag{9.18}$$

Also, the effective reactance of the transformer referred to the primary winding is

$$X_{E1} = X_1 + X_2' \tag{9.19}$$

where X_2' is the *reactance of the secondary winding referred to the primary winding.* For X_2' to produce the same effect in the primary winding as X_2 in the secondary winding, each must absorb the same number of reactive volt amperes. The voltage across X_2' when carrying I_1 is $I_1 X_2'$, and the voltage across X_2 when carrying I_2 is $I_2 X_2$. Equating the reactive volt amperes consumed by the two elements gives

$$I_1{}^2 X_2' = I_2{}^2 X_2$$

or

$$X_2' = X_2 (I_2/I_1)^2 = X_2 (N_1/N_2)^2 \tag{9.20}$$

hence

$$X_{E1} = X_1 + X_2 (N_1/N_2)^2 \tag{9.21}$$

A similar argument can be advanced when referring the values of the primary winding of the transformer to the secondary (see figure 9.11b), in which case

$$\left. \begin{array}{l} R_{E2} = R_2 + R_1' = R_2 + R_1 (N_2/N_1)^2 \\ X_{E2} = X_2 + X_1' = X_2 + X_1 (N_2/N_1)^2 \end{array} \right\} \tag{9.22}$$

Example 9.6

A single-phase transformer with a voltage step-down ratio of 3.3 kV/415 V has primary and secondary winding resistances of 0.8 Ω and 0.0125 Ω respectively, the corresponding leakage reactances being 4 Ω and 0.05 Ω. If the load is equivalent to a coil of resistance 5 Ω and inductive reactance 3.75 Ω, determine the value of the secondary winding terminal voltage and the power consumed by the load.

Solution

Since the results required above refer to the secondary winding, we use the simplified equivalent circuit in figure 9.12, in which

$$R_{E2} = R_2 + R_1(N_2/N_1)^2 = 0.0125 + 0.8(415/3300)^2$$
$$= 0.0251 \ \Omega$$
$$X_{E2} = X_2 + X_1(N_2/N_1)^2 = 0.05 + 4(415/3300)^2$$
$$= 0.1132 \ \Omega$$

The total impedance in the secondary circuit is

$$Z = Z_{E2} + Z_L = (5 + R_{E2}) + j(3.75 + X_{E2}) = 5.0215 + j3.8632 \ \Omega$$
$$= 6.336\underline{/37.57°}\Omega$$

and

$$I_2 = 415\underline{/0°} / Z = 415\underline{/0°} / 6.336\underline{/37.57°} = 65.5\underline{/-37.57°} \ \text{A}$$

The value of the load impedance is

$$Z_L = 5 + j3.75 = 6.25\underline{/36.87°} \ \Omega$$

Therefore

$$V_2 = I_2 Z_L = 65.5\underline{/-37.57°} \times 6.25\underline{/36.87°}$$
$$= 409.4\underline{/-0.7°} \ \text{V}$$

The power consumed by the load is

$$P_L = I_2{}^2 R_L = 65.5^2 \times 5 = 21\ 451 \ \text{W} = 21.51 \quad \text{kW}$$

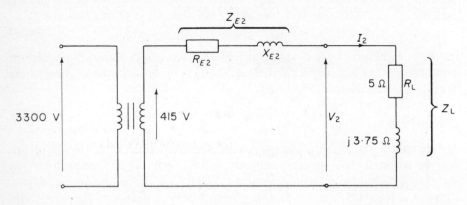

Figure 9.12

9.11 Per-unit Voltage Regulation

The per-unit voltage regulation of a transformer is the variation in the secondary voltage between no-load and full-load at a given power factor expressed as a proportion of the *no-load* secondary voltage, the primary voltage being meanwhile maintained at the rated voltage.

$$\text{Per-unit voltage regulation} = \frac{\text{no-load output voltage} - \text{full-load output voltage}}{\text{no-load output voltage}}$$

(9.23)

$$= \frac{V_1(N_2/N_1) - V_2}{V_1(N_2/N_1)} = \frac{V_1 - V_2(N_1/N_2)}{V_1}$$

(9.24)

The results of equations 9.23 and 9.24 are in a per-unit (p.u.) or dimensionless form. Should the results be required in a per cent form, the p.u. result is multiplied by 100. The per-unit voltage regulation of the transformer in example 9.6 is

$$\text{p.u. voltage regulation} = \frac{3300(415/3300) - 409.4}{3300(415/3300)}$$

$$= \frac{415 - 409.4}{415} = 0.0135 \text{ p.u. or } 1.35 \text{ per cent}$$

The per-unit voltage regulation for a lagging load is computed in terms of the parameters on the phasor diagram in figure 9.13 for the secondary winding of a transformer, as follows

$$E_2' = \sqrt{[(V_2 + I_2 R_{E2} \cos \phi + I_2 X_{E2} \sin \phi)^2 + (I_2 X_{E2} \cos \phi - I_2 R_{E2} \sin \phi)^2]}$$

(9.25)

From equation 9.23, the per-unit voltage regulation is

$$\text{per-unit voltage regulation} = (E_2' - V_2)/E_2'$$

(9.26)

The exact expression for E_2' may be simplified in many cases since the quadrature terms in equation 9.25 are small in value when compared with the in-phase terms (see example 9.7). Hence

$$E_2' \approx V_2 + I_2 R_{E2} \cos \phi + I_2 X_{E2} \sin \phi$$

(9.27)

and

$$\text{per-unit voltage regulation} = \frac{I_2 R_{E2} \cos \phi + I_2 X_{E2} \sin \phi}{E_2'}$$

(9.28)

The equivalent expression in terms of the primary winding quantities is

$$\text{per-unit voltage regulation} = \frac{I_1 R_{E1} \cos \phi + I_1 X_{E1} \sin \phi}{V_1}$$

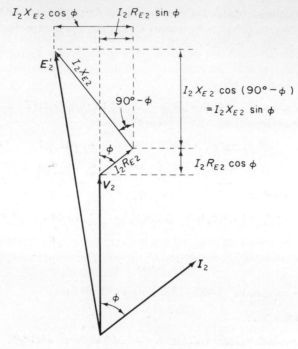

Figure 9.13   Determination of the voltage regulation of a transformer

Note For a load with a leading power factor, the sign associated with $I_2 X_{E2}$ in equations 9.25, 9.27, and 9.28 becomes negative,

The load phase angle which gives maximum voltage regulation can be determined by differentiating equation 9.28 with respect to ϕ and equating the result to zero. Its value is found to be a lagging phase angle which is equal to the angle in the internal impedance triangle of the transformer, that is, $\phi = \tan^{-1}(X_{E2}/R_{E2})$. For leading loads the voltage regulation first reduces, then becomes zero and finally becomes negative when the output voltage rises above the no-load voltage. For zero regulation to occur the numerator of equation 9.28 is zero, and occurs when

$$I_2 R_{E2} \cos \phi + I_2 X_{E2} \sin \phi = 0$$

or when

$$\phi = \tan^{-1}(-R_{E2}/X_{E2})$$

that is, it occurs when the load is capacitive.

Example 9.7

A transformer with resistance and reactance values referred to the secondary winding of $0.512\ \Omega$ and $1.173\ \Omega$, respectively, supplies a full load current of 12.5 A at a lagging power factor of 0.8. If the no-load secondary voltage is 400 V, determine the per-unit voltage regulation by means of (a) the exact expression and (b) the approximate expression, equation 9.28.

Solution

From the data given, $E_2' = 400$ V, $R_{E2} = 0.512$ Ω, $X_{E2} = 1.173$ Ω, $I_2 = 12.5$ A, $\cos \phi_2 = 0.8$ (therefore $\sin \phi_2 = 0.6$). From the above

$$R_{E2} \cos \phi_2 = 0.409 \quad R_{E2} \sin \phi_2 = 0.307$$

$$X_{E2} \cos \phi_2 = 0.938 \quad X_{E2} \sin \phi_2 = 0.704$$

(a) From equation 9.25

$$400^2 = [V_2 + 12.5(0.409 + 0.704)]^2 + [12.5(0.938 - 0.307)]^2$$

or

$$V_2 = 385.9 \text{ V}$$

per-unit voltage regulation = $(400 - 385.9)/400 = 0.035$ p.u.

(b) From equation 9.28

$$\text{per-unit voltage regulation} = 12.5(0.409 + 0.704)/400$$

$$= 0.035 \text{ p.u.}$$

9.12 Per-unit Resistance and Leakage Reactance Voltage Drops

It is sometimes convenient to express the full-load voltage drops occurring in a transformer as a fraction (usually in per unit or per cent) of the no-load terminal voltage. If

$$I_{FL1} = \text{nominal full-load primary current}$$

$$I_{FL2} = \text{nominal full-load secondary current}$$

$$V_1 = \text{nominal primary voltage}$$

$$E_2' = \text{nominal no-load secondary voltage}$$

then

$$\left.\begin{array}{l} \text{per-unit resistance drop} = I_{FL1}R_{E1}/V_1 \\ = I_{FL2}R_{E2}/E_2' \end{array}\right\} \quad (9.29)$$

and

$$\left.\begin{array}{l} \text{per-unit reactance drop} = I_{FL1}X_{E1}/V_1 \\ = I_{FL2}R_{E2}/E_2' \end{array}\right\} \quad (9.30)$$

Example 9.8

Calculate the values of the per-unit resistance and reactance drops in example 9.7.

Solution

From the data given in the problem, $I_{FL2} = 12.5$ A, $R_{E2} = 0.512$ Ω, $X_{E2} = 1.173$ Ω, $E_2' = 400$ V. From equation 9.29

per-unit resistance drop $= 12.5 \times 0.512/400 = 0.016$ p.u.

and from equation 9.30

per-unit reactance drop $= 12.5 \times 1.173/400 = 0.037$ p.u.

9.13 Transformer Efficiency

The per-unit efficiency of a transformer is given by the expression

$$\left.\begin{aligned}
\eta &= \frac{\text{output power}}{\text{input power}} = \frac{\text{output power}}{\text{output power} + \text{losses}} \\
&= \frac{\text{input power} - \text{losses}}{\text{input power}} = 1 - \frac{\text{losses}}{\text{input power}}
\end{aligned}\right\} \tag{9.31}$$

The power losses in the transformer are divided into two main groups, namely

    (a) losses which vary with the load current
    (b) losses which vary with the core flux

The first group consists of the *copper losses*, P_c, which for a two-winding transformer are

$$P_c = I_1{}^2 R_1 + I_2{}^2 R_2 = I_1{}^2 R_{E1} = I_2{}^2 R_{E2} \quad \text{watts} \tag{9.32}$$

The second group of losses, group b above, can be further subdivided into the *hysteresis loss*, P_h, and the *eddy current loss*, P_e. The hysteresis loss, discussed in section 3.14, is given by the relationship

$$P_h = kfB_m{}^n \quad \text{watts/m}^3$$

where k is a constant, f is the supply frequency, B_m is the maximum value of flux density in the core, and n is a number whose value lies between 1.6 and 2.

The eddy current loss is due to flow of eddy currents in the magnetic material of the core. This loss is reduced to an economic minimum value by using a laminated core of high resistivity material. This loss is given by the expression

$$P_e = Kf^2 B_m{}^2 \quad \text{watts/m}^3$$

where K is a constant, and f and B_m are defined above.

The core loss, P_0, is

$$P_0 = P_h + P_e$$

Since both f and B_m are constant for a given supply frequency and transformer, the iron loss is approximately constant over the working load range of the transformer. The value of P_0 is determined by measuring the no-load power consumed by the transformer (see also sections 9.6 and 9.16).

All-day efficiency

In many instances the load connected to the transformer varies over a 24-hour period, and its efficiency will vary from a low value under light-load conditions to a high value at higher values of load. In these applications the instantaneous efficiency expressed by equation 9.31 may be less important that its *all-day efficiency*, which is the ratio of the output energy to the input energy of a 24-hour period.

$$\text{All-day efficiency} = \frac{\text{output energy during 24 hours}}{\text{input energy during 24 hours}}$$

9.14 Conditions for Maximum Efficiency

The per-unit efficiency at load current I_2 is

$$\eta = \frac{\text{output power}}{\text{input power}} = \frac{V_2 I_2 \cos \phi_2}{V_2 I_2 \cos \phi_2 + I_2{}^2 R_{E2} + P_0}$$

$$= \frac{V_2 \cos \phi_2}{V_2 \cos \phi_2 + I_2 R_{E2} + P_0/I_2} \qquad (9.33)$$

where $I_2{}^2 R_{E2}$ is the total copper loss of the transformer referred to the secondary winding. Maximum efficiency occurs when $d\eta/dI_2 = 0$, and the condition for its occurrence is deduced by differentiating equation 9.33 with respect to I_2. Since V_2 and $\cos \phi_2$ are constant for a given load, the condition for maximum efficiency occurs when the denominator of equation 9.33 is a minimum, that is, when

$$d(V_2 \cos \phi_2 + I_2 R_{E2} + P_0/I_2)/dI_2 = 0$$

or when

$$R_{E2} - P_0/I_2{}^2 = 0 \qquad (9.34)$$

that is

$$I_2{}^2 R_{E2} = P_0$$

that is, when

$$\text{copper losses} = \text{iron losses} \qquad (9.35)$$

To verify that expression 9.35 gives the minimum value to the denominator of equation 9.33, the expression on the left-hand side of equation 9.34 is differentiated with respect to I_2, giving a solution of $2P_0/I_2{}^2$. Since this has a positive value, then expression 9.35 is the condition for maximum efficiency to occur.

9.15 Effect of Load Current on the Transformer Copper Loss

The transformer copper losses are proportional to (current)^2, and doubling the load current causes the copper losses to quadruple.

Since the supply voltage remains approximately constant, then

$$I^2 \propto (VI)^2 \qquad (9.36)$$

hence

$$\text{copper losses} \propto (\text{volt amperes consumed by the load})^2 \qquad (9.37)$$

Example 9.9

A 40-kVA, 4000/400-V single-phase transformer has a core loss of 450 W and a full-load copper loss of 850 W. For a power factor of 0.8 lagging, calculate (a) the full-load efficiency, (b) the maximum efficiency and the value of primary current at which it occurs. Determine (c) the efficiency of the transformer when supplying a load of 20 kVA at 0.8 power factor lagging.

Solution

(a) Total losses at full load = 450 + 850 = 1300 W = 1.3 kW

$$\text{Output power} = 40 \times 0.8 = 32 \text{ kW}$$

Hence

$$\eta = 1 - \frac{\text{losses}}{\text{input power}} = 1 - \frac{1.3}{32 + 1.3} = 1 - 0.039 = 0.961 \text{ p.u.}$$

(b) For maximum efficiency, copper loss = iron loss, hence

$$\text{total loss} = 2 \times \text{iron loss} = 900 \text{ W}$$

From equation 9.37

$$\frac{\text{copper loss at maximum efficiency}}{\text{copper loss at full load}} = \left(\frac{\text{VA for maximum efficiency}}{\text{VA for full load}}\right)^2$$

$$= \left(\frac{\text{current for maximum efficiency}}{\text{full-load current}}\right)^2$$

therefore

$$\text{current for maximum efficiency} = I_{FL}\ \left(\frac{\text{copper loss at maximum efficiency}}{\text{copper loss at full load}}\right)$$

and the full-load primary current is

$$I_{FL} = 40\ 000/4000 = 10\ \text{A}$$

hence the primary current which gives maximum efficiency is

$$I_{FL}\sqrt{(450/850)} = 7.28\ \text{A}$$

The input power at this current is

$$V_1 I_1 \cos\phi = 4000 \times 7.28 \times 0.8 = 23\ 300\ \text{W} = 23.3\ \text{kW}$$

and the value of the maximum efficiency is

$$\eta = 1 - \frac{\text{losses}}{\text{input power}} = 1 - \frac{0.9}{23.3}$$

$$= 0.9614\ \text{p.u.}$$

(c) From equation 9.37

$$\frac{\text{full-load copper loss}}{\text{copper loss at 20 kVA}} = \left(\frac{40}{20}\right)^2 = 4$$

Hence

$$\text{copper loss at 20 kVA} = 850/4 = 212.5\ \text{W} = 0.2125\ \text{kW}$$

At this load the output power is $20 \times 0.8 = 16\ \text{kW}$

$$\text{Efficiency} = 1 - \frac{\text{losses}}{\text{input power}} = 1 - \frac{0.45 + 0.2125}{16 + 0.45 + 0.2125}$$

$$= 0.9602\ \text{p.u.}$$

9.16 Open-circuit and Short-circuit Tests

In order to determine the values of the parameters R_C and X_m in figure 9.9, and R_E and X_E in figure 9.11, tests must be carried out on the transformer. The open-circuit and the short-circuit tests described below enable these parameters to be evaluated, the results also permitting the efficiency and the voltage regulation to be computed.

Open-circuit test

The open-circuit test (or no-load test) was discussed in section 9.6 in connection with the determination of the parameters R_C and X_m. The connections for the test

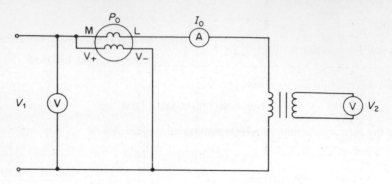

Figure 9.14   The open-circuit test

are shown in figure 9.14, in which the transformer is energised at its rated voltage, V_1, and the values of the input current I_0 and input power, P_0, together with the value of the secondary voltage V_2 are measured. To reduce the possibility of errors in the measurements, the primary winding voltmeter is connected on the 'mains' side of the wattmeter, and the ammeter is on the 'load' side. These connections prevent the power consumed by the voltmeter being measured, and also prevent the current drawn by the wattmeter voltage coil being measured by the ammeter. The open-circuit power factor is

$$\cos \phi_0 = P_0/V_1 I_0$$

and

$$I_C = I_0 \cos \phi_0 \quad \text{also} \quad I_{\text{mag}} = I_0 \sin \phi_0$$

therefore

$$R_C = V_1/I_C = V_1/I_0 \cos \phi_0 = V_1/(P_0/V_1) = V_1{}^2/P_0 \tag{9.38}$$

and

$$X_m = V_1/I_{\text{mag}} = V_1/I_0 \sin \phi_0 \tag{9.39}$$

Short-circuit test

In the short-circuit test, figure 9.15, one winding is short-circuited and the voltage applied to the other winding is increased from zero until full-load current flows. Readings of the short-circuit current, $I_{1\text{SC}}$, the primary voltage, $V_{1\text{SC}}$, and the input power, P_{SC}, are noted. The connections shown in figure 9.15 are used to minimise sources of error in measuring the quantities concerned. Since the excitation voltage is low (typically about 10 per cent of the nominal supply voltage), the core losses are small and the reading P_{SC} of the wattmeter is taken to be equal to the full-load copper loss. The short-circuit power factor is

$$\cos \phi_{\text{SC}} = P_{\text{SC}}/(V_{1\text{SC}} I_{1\text{SC}}) \tag{9.40}$$

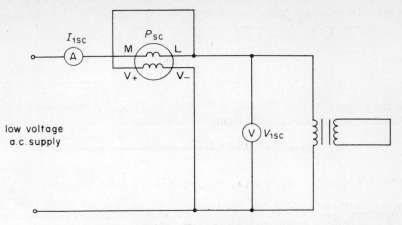

Figure 9.15 The short-circuit test

and the impedance between the primary terminals is

$$Z_{E1} = \frac{V_{1SC}}{I_{1SC}} = Z_{E1}/\phi_{SC} = R_{E1} + jX_{E1} \qquad (9.41)$$

The values of the primary and the secondary winding resistances can be separated by d.c. tests on the individual windings, but it is impossible to separate the values of the primary and secondary leakage reactances.

9.17 Determination of the Efficiency and the Voltage Regulation from the Open-circuit and Short-circuit Tests

Efficiency

$$\text{Total losses} = \text{copper loss} + \text{iron loss} = P_{SC} + P_0$$

At a load power factor $\cos \phi$, the full-load efficiency is

$$\frac{\text{rated VA} \times \cos \phi}{(\text{rated VA} \times \cos \phi) + P_{SC} + P_0} \quad \text{p.u.} \qquad (9.42)$$

For any other VA consumption, say $(VA)_2$, the efficiency is

$$\frac{(VA)_2 \times \cos \phi}{[(VA)_2 \cos \phi] + \{ [(VA)_2/\text{rated VA}]^2 \times P_{SC} \} + P_0} \qquad (9.43)$$

Voltage regulation

From equation 9.41, the magnitude of the impedance of the transformer referred to the primary winding is

$$Z_{E1} = V_{1SC}/I_{1SC}$$

and the total resistance referred to the primary winding is

$$R_{E1} = P_{SC}/I_{1SC}^2$$

hence

$$X_{E1} = \sqrt{(Z_{E1}^2 - R_{E1}^2)}$$

and the primary winding full-load current is $I_1 = I_{1SC}$. From the work on voltage regulation, for a current I_1 at a lagging phase angle ϕ

$$\text{p.u. voltage regulation} = (I_1 R_{E1} \cos \phi + I_1 X_{E1} \sin \phi)/V_1 \qquad (9.44)$$

Example 9.10

The results of open-circuit and short-circuit tests carried out on the same winding of a 3.5 kVA single phase transformer are as follows

  open-circuit: current 1.2 A, voltage 200 V, power 24 W
  short-circuit: current 17.5 A, voltage 6.4 V, power 28 W

Determine, for a primary voltage of 200 V and a load power factor of 0.8 lagging (a) the efficiency of the transformer at full-load, and (b) the per-unit voltage regulation.

Solution

(a) From equation 9.42

$$\eta = \text{rated VA} \times \cos \phi/[(\text{rated VA} \times \cos \phi) + P_{SC} + P_0]$$

$$= 3500 \times 0.8/[(3500 \times 0.8) + 28 + 24] = 0.982 \text{ p.u.}$$

(b)

$$Z_{E1} = V_{1SC}/I_{1SC} = 6.4/17.5 = 0.366 \ \Omega$$

$$R_{E1} = P_{SC}/I_{1SC}^2 = 28/17.5^2 = 0.091 \ \Omega$$

$$X_{E1} = \sqrt{(Z_{E1}^2 - R_{E1}^2)} = \sqrt{(0.366^2 - 0.091^2)} = 0.354 \ \Omega$$

The full-load primary current is $3500/200 = 17.5$ A, hence

$$\text{p.u. voltage regulation} = I_1(R_{E1} \cos \phi + X_{E1} \sin \phi)/V_1$$

$$= 17.5[(0.091 \times 0.8) + (0.354 \times 0.6)]/200$$

$$= 0.025 \text{ p.u.}$$

9.18 Transformer Construction

The preceding work on the transformer equivalent circuit has illustrated the desirability of reducing the leakage reactance of the transformer. The principal

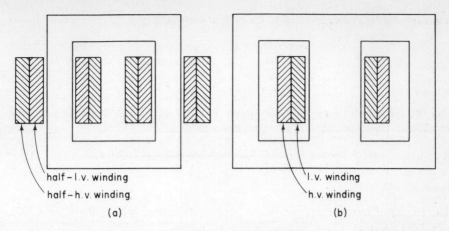

half – l.v. winding
half – h.v.winding
(a)

l.v. winding
h.v. winding
(b)

Figure 9.16   (a) Core type construction and (b) shell type construction

types of magnetic circuit and winding arrangements used in power transformers are described below.

Magnetic circuit

The principal types of magnetic circuit in use are the *core type* and the *shell type*, illustrated in figures 9.16a and b, respectively. In the core type, one-half of each winding is associated with each limb, the magnetic circuit having a uniform cross-sectional area. In the shell type, both windings are on the centre limb, which has twice the cross-sectional area of each of the outer limbs. The object of both types of core construction is to place the windings into intimate contact with one another so as to reduce the leakage flux.

Windings

The winding arrangements used in power transformers generally take one of the two forms in figure 9.17. In the *concentric construction*, figure 9.17a, the low voltage winding is placed nearer to the iron core. In the *sandwich construction*, figure 9.17b, the high voltage winding is sandwiched between the two halves of the low voltage winding.

In large transformers, and also in some small ones, ventilation spaces are left in the windings to allow circulation space for the coolant, which is usually either air or oil.

9.19 Auto-transformers

An auto-transformer has a single winding, part of which is common to both the primary and secondary circuits, as shown in figure 9.18. As with all transformers, ampere-turn balance is maintained between the windings, and the number of volts

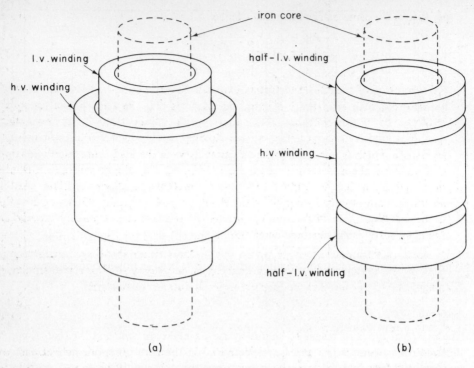

Figure 9.17   (a) Concentric winding construction and (b) sandwich winding construction

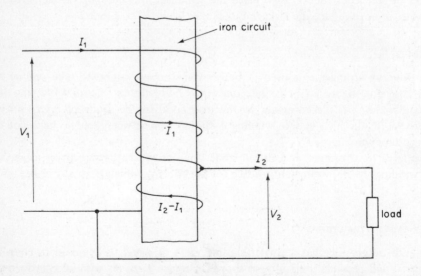

Figure 9.18   The auto-transformer

per turn is the same on each winding. Hence

$$\frac{V_2}{V_1} = \frac{N_2}{N_1} = \frac{I_1}{I_2} \qquad\qquad (9.45)$$

An advantage of the auto-transformer over the double-wound transformer, that is, a transformer with electrically isolated windings, is that the value of the secondary current has a lower value than that which would flow in an equivalent double-wound transformer. The current flowing in the lower half of the winding is $(I_2 - I_1)$, which is less than current I_2 that flows in the load. This results not only in a saving in the amount of copper required in winding the transformer, but also in a reduction in the copper loss and an increase in efficiency. The greatest advantages are obtained when the ratio V_2/V_1 approaches unity. Offset against this is the fact that in auto-transformer circuits the primary and secondary circuits are not electrically isolated from one another.

Auto-transformers are used where the features listed above are advantageous, that is, in systems where the primary and secondary voltages are similar in magnitude to one another, and for use in induction motor starters.

9.20 Current Transformers

When measuring large values of alternating current, it is more convenient to 'transform' the magnitude of the current to a low value, in the range 0–5 A, than it is to use an instrument specially designed to measure the current directly.

One form of current transformer (C.T.) is shown in figure 9.19, in which the

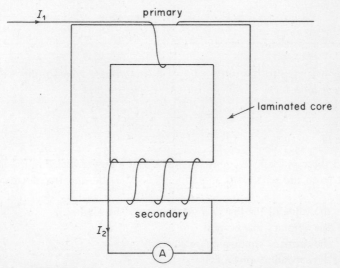

Figure 9.19   A basic form of current transformer

primary winding either consists of a few turns of very heavy wire or is simply a conductor passing through the centre of the magnetic circuit. The secondary circuit consists of many turns of fine wire, the secondary current rating usually being 5 A. In operation, the C.T. secondary winding should either have an ammeter connected between its terminals or it should be short-circuited. If the secondary becomes open-circuited, ampere-turn balance on the core is no longer maintained and the core flux rises to a very high value. The consequence is that a dangerously high voltage is induced in the secondary winding, and also that the iron circuit dissipates a considerable amount of energy (resulting from the high value of core flux) and becomes very hot.

9.21 Polyphase Transformers

The majority of polyphase power transformers employ a core type of magnetic circuit construction, as shown in figure 9.20a. The transformer shown needs only three limbs, and is more economic in use than three separate single-phase transformers. Many different types of winding interconnection are possible and, in the case shown, the primary winding is delta-connected and the secondary is star-connected − see also figure 9.20b.

The general theory of the transformer outlined in earlier sections is applicable to three-phase transformers if we note that, in general, the relationships apply to *each limb* (that is, each phase) of the transformer. Thus, for limb 1 of the transformer in figure 9.20a

$$\frac{V_{P1}}{V_{P2}} = \frac{N_1}{N_2} = \frac{I_{P2}}{I_{P1}} \tag{9.46}$$

Thus, when referring to the *turns ratio* of a three-phase transformer we mean the *turns ratio per limb*.

Example 9.11

A balanced three-phase load consumes 500 kW at a power factor of 0.8 lagging, and is supplied from the secondary of a 33 kV/11 kV delta−star transformer. Calculate

(a)   the value of the line current drawn by the load
(b)   the primary line current
(c)   the primary winding phase current
(d)   the secondary phase voltage
(e)   the turns ratio on each limb of the transformer

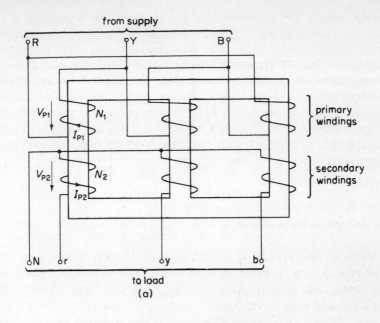

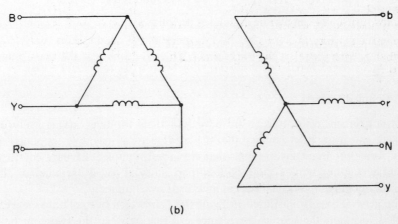

(b)

Figure 9.20 A three-phase core-type delta–star transformer

Solution

If V_{L2} and I_{L2} are the values of the secondary line voltage and current, respectively, then

(a)

$$P = \sqrt{3}\ V_{L2} I_{L2} \cos \phi$$

or

$$I_{L2} = P/\sqrt{3}\ V_{L2}\cos\phi = 500\ 000/(\sqrt{3}\times 11000\times 0.8) = 32.8\ \text{A}$$

(b) Assuming the transformer to be 100 per cent efficient, then the apparent power consumed by the load is equal to the apparent power delivered to the primary winding.

$$\sqrt{3}\ V_{L1}I_{L1} = \sqrt{3}\ V_{L2}I_{L2}$$

or

$$I_{L1} = V_{L2}I_{L1}/V_{L1} = 11000\times 32.8/33000 = 10.93\ \text{A}$$

where V_{L1} and I_{L1} are the values of the primary line voltage and current, respectively.

(c) The primary winding phase current is

$$I_{P1} = I_{L1}/\sqrt{3} = 10.93/\sqrt{3} = 6.31\ \text{A}$$

(d) If the secondary phase voltage is V_{P2}, then

$$V_{P2} = V_{L2}/\sqrt{3} = 11000/\sqrt{3} = 6351\ \text{V}$$

(e) Since the primary winding is delta-connected, then $V_{P1} = V_{L1} = 33000$ V. From equation 9.46

$$\text{turns ratio} = N_1/N_2 = V_{P1}/V_{P2} = 33000/6351 = 5.196{:}1$$

Note the turns ratio can also be calculated as follows

$$N_1/N_2 = I_{P2}/I_{P1} = I_{L2}/I_{P1} = 32.8/6.31$$

$$= 5.196{:}1$$

9.22 Coupled Circuits

The transformers considered so far have had their windings magnetically closely coupled, and it has been assumed that the coefficient of magnetic coupling is unity. Many forms of circuit involve coils that are loosely coupled, having a coefficient of magnetic coupling with a value of less ·than unity. Typical examples of this are found in radio, television, and telecommunications equipment.

The general case is illustrated in figure 9.21, in which two coils are magnetically coupled together, the mutual inductance existing between the two coils being *M*.

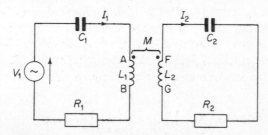

Figure 9.21   Magnetically coupled circuits

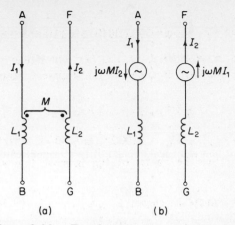

Figure 9.22   E.m.f.s induced in coupled circuits

With the coupling shown, when current enters terminal A of L_1, the induced e.m.f. in coil L_2 is such as to make point F instantaneously positive with respect to point G, and current I_2 flows out of terminal F. The circuit is analysed by applying the principle of the 'dot' notation outlined in chapter 3.

If we consider coils L_1 and L_2 independently, shown in figure 9.22a, it is found to be possible to replace the mutual coupling between the two coils by a voltage generator in series with each coil, the magnitude of each generator being

$$j\omega M \times \text{(current producing the induced e.m.f.)}$$

Thus, a voltage generator of magnitude $j\omega MI_2$ is included in series with winding L_1, and a generator of magnitude $j\omega MI_1$ is included in series with winding L_2. The direction in which the induced e.m.f. acts is deduced by applying the dot notation principle as follows. According to this notation, if a current enters a terminal marked with a 'dot' then the induced e.m.f. in all the coupled coils is such that it causes the ends of those coils marked with dots to have an instantaneous positive polarity. If current leaves a terminal marked with a 'dot', then the ends of the other coupled coils marked with dots become instantaneously negative.

Since I_1 enters terminal A (marked with a dot), then the induced e.m.f. $j\omega MI_1$ in coil L_2 causes point F (also marked with a dot) to be positive with respect to point G. Also, since I_2 leaves terminal F, then the induced e.m.f. $j\omega MI_2$ in coil L_1 causes terminal A on the primary winding to be negative with respect to terminal B. The resulting directions of induced e.m.f. are as shown in figure 9.22b. The completed equivalent circuit of figure 9.21 is shown in figure 9.23.

The mesh equation for the primary circuit of figure 9.23 is

$$V_1 = I_1\left[R_1 + j\left(\omega L_1 - \frac{1}{\omega C_1}\right)\right] - j\omega MI_2 \qquad (9.47)$$

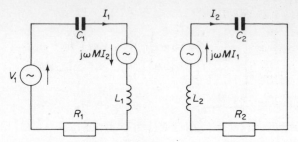

Figure 9.23 Equivalent electrical circuits of two magnetically coupled circuits

$$= I_1 Z_{11} - j\omega M I_2 \tag{9.48}$$

and for the secondary winding is

$$0 = -j\omega M I_1 + I_2 \left[R_2 + j \left(\omega L_2 - \frac{1}{\omega C_2} \right) \right] \tag{9.49}$$

$$= -j\omega M I_1 + I_2 Z_{22} \tag{9.50}$$

where Z_{11} and Z_{22} are described as the *self-impedances* of the primary and secondary meshes, respectively. Solving for I_2 from equation 9.50 yields

$$I_2 = j\omega M I_1 / Z_{22} \tag{9.51}$$

and for V_1 between equations 9.48 and 9.50 gives

$$V_1 = I_1 (Z_{11} + \omega^2 M^2 / Z_{22}) \tag{9.52}$$

Example 9.12

A circuit of the type in figure 9.21 is energised by a supply of 100 V at a frequency of $10^6/2\pi$ Hz, the circuit values being $R_1 = 20\ \Omega$, $L_1 = 200\ \mu\text{H}$, $C_1 = 1.25\ \text{nF}$, $R_2 = 10\ \Omega$, $L_2 = 100\ \mu\text{H}$, $C_2 = 2.5\ \text{nF}$, and $M = 75\ \mu\text{H}$. Determine the effective values of primary resistance and reactance, and also calculate the values of the primary and secondary currents. Evaluate also the coefficient of magnetic coupling between the coils.

Solution

From equations 9.47 and 9.48

$$Z_{11} = R_1 + j \left(\omega L_1 - \frac{1}{\omega C_1} \right) = 20 + j \left[(10^6 \times 200 \times 10^{-6}) - \frac{1}{10^6 \times 1.25 \times 10^{-9}} \right]$$

$$= 20 - j600\ \Omega$$

From equations 9.49 and 9.50

$$Z_{22} = R_2 + j\left(\omega L_2 - \frac{1}{\omega C_2}\right) = 10 + j\left[(10^6 \times 100 \times 10^{-6}) - \frac{1}{10^6 \times 2.5 \times 10^{-9}}\right]$$

$$= 10 - j300 \ \Omega$$

Also, from equation 9.52

$$\text{input impedance} = V_1/I_1 = Z_{11} + \omega^2 M^2/Z_{22}$$

$$= (20 - j600) + (10^6)^2 (75 \times 10^{-6})^2/(10 - j300)$$

$$= 20.63 - j581.25$$

Hence

$$\text{effective primary resistance} = 20.63 \ \Omega$$
$$\text{effective primary reactance} = -j581.25 \ \Omega$$

Current I_1 is computed from equation 9.52

$$I_1 = V_1/\text{primary input impedance}$$

$$= 100/(20.63 - j581.25) \approx j0.172 \text{ A} = 0.172 \ \underline{/90°} \text{ A}$$

and from equation 9.50

$$I_2 = j\omega M I_1/Z_{22} = j10^6 \times 75 \times 10^{-6} \times j0.172/(10 - j300)$$

$$\approx -j0.0427 \text{ A} = 0.0427 \ \underline{/-90°} \text{ A}$$

and the coupling coefficient k is

$$k = M/\sqrt{(L_1 L_2)} = 75 \times 10^{-6}/\sqrt{(200 \times 10^{-6} \times 100 \times 10^{-6})}$$

$$= 0.53$$

Summary of essential formulae

General relationships: $\quad \dfrac{V_2}{V_1} = \dfrac{N_2}{N_1} = \dfrac{I_1}{I_2}$

Input impedance with a resistive load: $\quad R_1 = R_L(N_1/N_2)^2$

E.M.F. equation: $\quad E_1 = 4.44 \, fN_1 \, \Phi_m$

$$E_2 = 4.44 \, fN_2 \, \Phi_m$$

No-load current: $\quad I_0 = I_C + I_{mag} = I_0 \ \underline{/\phi_0}$

$$\text{where} \quad I_0 = \sqrt{(I_C^2 + I_{mag}^2)} \quad \text{and} \quad \phi_0 = \tan^{-1}(I_{mag}/I_C)$$

Equivalent resistance, referred to primary: $\quad R_{E1} = R_1 + R_2(N_1/N_2)^2$

referred to secondary: $\quad R_{E2} = R_2 + R_1(N_2/N_1)^2$

Equivalent leakage reactance, referred to primary: $X_{E1} = X_1 + X_2(N_1/N_2)^2$

referred to secondary: $X_{E2} = X_2 + X_1(N_2/N_1)^2$

Per-unit voltage regulation:

$$\text{p.u. regulation} = \frac{\text{no-load output voltage} - \text{full-load output voltage}}{\text{no-load output voltage}}$$

$$\approx \frac{I_2(R_{E2} \cos \phi + X_{E2} \sin \phi)}{\text{no-load output voltage}}$$

Per-unit efficiency: η = power output/power input

$$= V_2 I_2 \cos \phi_2 / (V_2 I_2 \cos \phi_2 + I_2{}^2 R_{E2} + P_0)$$

$$\text{all-day efficiency} = \frac{\text{energy output during 24 hours}}{\text{energy input during 24 hours}}$$

Open-circuit test: $\cos \phi_0 = P_0 / V_1 I_0$

$$I_C = I_0 \cos \phi_0$$

$$I_{\text{mag}} = I_0 \sin \phi_0$$

$$R_C = V_1{}^2 / P_0$$

$$X_m = V_1 / I_{\text{mag}}$$

Short-circuit test: $Z_{E1} = V_{1SC}/I_{1SC}$

$$R_{E1} = P_{SC}/I_{1SC}{}^2$$

$$X_{E1} = \sqrt{(Z_{E1}{}^2 - R_{E1}{}^2)}$$

Efficiency of a transformer providing an output of $(VA)_2$ at a power factor of $\cos \phi$:

$$\eta = (VA)_2 \cos \phi \left/ \left[(VA)_2 \cos \phi + \left(\frac{(VA)_2}{\text{rated VA}} \right)^2 P_{SC} + P_0 \right] \right.$$

Polyphase transformer, per limb (or phase): $\dfrac{V_{P1}}{V_{P2}} = \dfrac{N_1}{N_2} = \dfrac{I_{P2}}{I_{P1}}$

Coupled circuit, mutually induced e.m.f.: $E_2 = j\omega M I_1$

PROBLEMS

9.1 An ideal transformer has a step-up voltage ratio of 1:2. Calculate the value of the impedance measured between the primary terminals when a resistance of 2000 Ω is connected between the secondary terminals. How is this value modified at a frequency of 200 Hz if the inductance of the primary winding is 1.0 H?
[500 Ω; 464.5 Ω]

9.2 The numbers of turns on the primary and secondary windings, respectively, of a single-phase transformer are 605 and 55. If the primary winding is supplied at 3.3 kV, determine (a) the no-load secondary winding voltage, (b) the primary winding current and its power factor when the secondary winding current is 300 A at 0.8 power factor lagging, the no-load primary current being 7.5 A at 0.2 power factor lagging. Draw to scale the phasor diagram for the primary and secondary circuits.

[(a) 300 V; (b) 33.25 A at a power factor of 0.7011 lagging]

9.3 The following data refer to a single-phase transformer:

> peak flux density in the core = 1.2 T
> net core area = 0.04 m^2
> supply voltage = 2500 V, 50 Hz
> step-down voltage ratio = 5:1

The winding arrangement is shown in figure 9.24, where N_1 and N_2 refer respectively to the primary and secondary windings. If the number of turns on each section of the winding is to be given as an integer, determine the number of turns on each section of the winding. Explain the basis for the selection of the number of turns.

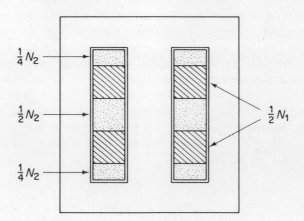

Figure 9.24

The no-load current drawn by the primary winding of the above transformer is 2 A at a power factor of 0.4 lagging. Calculate the magnitude and the power factor of the primary current when a load of 100 A at a lagging power factor of 0.8 is connected to the secondary terminals of the transformer. Neglect the effects of the power loss in the transformer and of the voltage drops in the transformer.

[N_1 = 240 turns, N_2 = 48 turns; 21.67 A at 0.772 p.f. lagging]

9.4 If the ratio of the secondary voltage to primary voltage of an auto-transformer is N (= V_2/V_1), show that the ratio of the copper required in an auto-transformer to that required in a two-winding transformer of the same rating is given by (1~N).

An auto-transformer is required to increase a supply voltage from 220 V r.m.s. to 250 V. Determine the position of the tapping point on the transformer winding as a percentage of the total number of turns on the transformer. Estimate the current in each part of the winding if the load is 10 kVA.

A 10-kV/2 kV two-winding transformer is rated at 100 kVA. If the two windings are connected in series with one another to form an auto-transformer, determine the two possible values of voltage ratio, and the kVA rating in each case.
[88%; 40 A, 5.45 A; 6:1, 120 kVA; 1.2:1, 600 kVA]

9.5 The primary and secondary windings of a 50-kVA, 6600/250-V, single-phase transformer have resistances of 12 Ω and 0.02 Ω, respectively. The leakage reactance referred to the h.v. winding is 35 Ω. Neglecting the effects of the no-load current, calculate (a) the voltage regulation at (i) unity power factor load, (ii) 0.8 lagging power factor load and (b) the primary voltage required to circulate full-load secondary current when the secondary terminals are short-circuited.
[(a) (i) 0.03 p.u., (ii) 0.0479 p.u.; (b) 330 V]

9.6 When a capacitive load is connected to the secondary terminals of a transformer which has a high value of leakage reactance, the secondary terminal voltage is found to rise above its no-load value. With the aid of a phasor diagram and appropriate analysis, explain the reason for this phenomenon.

9.7 Derive an expression for the equivalent resistance and equivalent reactance referred to the secondary winding of a transformer.

The h.v. and l.v. windings of a 4000/400-V, single-phase transformer have resistances of 1.3 Ω and 0.015 Ω, respectively. Determine the magnitude of the secondary terminals voltage when 4000 V is applied to the primary winding, and a 1-Ω resistor is connected to the l.v. terminals. Calculate the power consumed by the resistor.
[384.6 V; 147.92 kW]

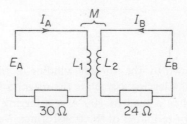

Figure 9.25

9.8 When $E_A = 100$ V at 50 Hz is applied to the circuit in figure 9.25, it is found that $I_A = 2$ A, $I_B = 0$ and $E_B = 54$ V. When the supply is disconnected from E_A and is connected to E_B, it is found that $I_B = 2.5$ A ($I_A = 0$). Determine the value of L_1, L_2 and M. Calculate also the value of the coupling coefficient and also the open-circuit e.m.f. E_A when the circuit is energised at E_B.
[0.127 H; 0.102 H; 0.086 H; 0.755; 67.6 V]

10 Transients in RL and RC Circuits

After a circuit is connected to an electrical supply, it takes a little time for the current and voltage associated with each component to settle down to their *steady-state* value. The length of this settling period is known as the *transient period* of operation, the transients dying away in a period of time known as the *settling-time*. In this chapter we are concerned with transients occurring in *RL* and *RC* circuits.

10.1 Transients in *RL* Series Circuits

Rise of current

Consider the inductive circuit in figure 10.1. When the switch is closed, the loop equation is

$$E = iR + L\frac{di}{dt} \tag{10.1}$$

where i is the instantaneous value of the circuit current. Equation 10.1 is the *differential equation* of the circuit, and its solution gives an expression for the instantaneous value of current. Equation 10.1 is rewritten as follows

$$\frac{R}{L}dt = \frac{di}{\frac{E}{R} - i} \tag{10.2}$$

so that

$$\int \frac{R}{L}dt = \int \frac{di}{\frac{E}{R} - i}$$

Integrating both sides of the equation yields

$$\frac{Rt}{L} = -\ln\dagger\left(\frac{E}{R} - i\right) + \ln A \tag{10.3}$$

†In is sometimes written $\log_e$, where e (2.718 28) is the base of natural logarithms.

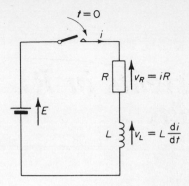

Figure 10.1   *RL* series circuit energised by a d.c. source

where $\ln A$ is the constant of integration, whose value is determined from a knowledge of the conditions in the circuit when $t = 0$, that is, the *initial conditions* in the circuit. If the initial value of the current is zero, that is, $i = 0$ when $t = 0$, then substituting these values into equation 10.3 gives

$$0 = - \ln \frac{E}{R} + \ln A$$

or

$$A = E/R$$

Substituting the above value of A into equation 10.3 yields

$$\frac{Rt}{L} = - \ln \left(\frac{E}{R} - i \right) + \ln \frac{E}{R} = \ln \left(\frac{E/R}{(E/R) - i} \right)$$

or

$$e^{Rt/L} = \frac{E/R}{(E/R) - i}$$

Solving for i gives

$$i = \frac{E}{R} (1 - e^{-Rt/L}) \tag{10.4}$$

The instantaneous voltage across resistor R is

$$v_R = iR = E(1 - e^{-Rt/L}) \tag{10.5}$$

and the instantaneous voltage across inductor L is

$$v_L = E - v_R = Ee^{-Rt/L} \tag{10.6}$$

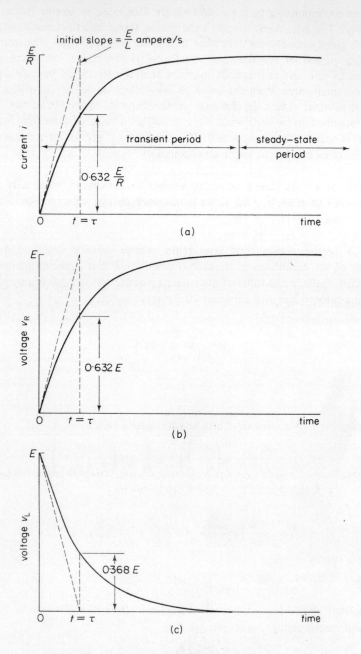

Figure 10.2   Rise of current in a *RL* series circuit

The curves corresponding to i, v_R, and v_L are illustrated in figures 10.2a, b and c, respectively. The equations contain exponential terms which, theoretically at any rate, reach steady conditions only after an infinite period of time. This implies that the settling period of the transient is infinite! However, we can apply practical limitations to this, and it is usually assumed that the transients have decayed to a sufficiently small value when the value of the voltage or current is within one per cent of their final values. In the case of the current, figure 10.2a, the transient period is assumed to be completed when i reaches a value of $0.99E/R$; for the v_R waveform it occurs when it reaches $0.99E$; and for v_L it occurs when it has fallen to $0.01E$. The three conditions occur simultaneously.

Final value of i As stated above, i reaches its maximum value only after an infinite length of time, and its value is obtained by inserting $t = \infty$ into equation 10.4 as follows:

$$(i)_{t=\infty} = \frac{E}{R}(1 - e^{-\infty}) = \frac{E}{R} \tag{10.7}$$

Initial rate of rise of current The initial rate of rise of current can be computed from equation 10.1 if we insert the value of the initial current, viz. $(i)_{t=0} = 0$, into the equation as follows:

$$E = R(i)_{t=0} + L\left(\frac{di}{dt}\right)_{t=0}$$

hence

$$\left(\frac{di}{dt}\right)_{t=0} = \frac{E}{L} \ \text{A/s}$$

If the initial rate of rise of current were maintained, it would reach the final value of current of E/R in τ seconds, see figure 10.2a, where

$$\tau = \frac{\text{final value}}{\text{initial rate of rise}} = \frac{E/R}{E/L} = \frac{L}{R} \ \text{seconds} \tag{10.8}$$

Parameter τ is known as the *time constant* of the circuit.

Circuit current when $t = \tau$ Equation 10.4 may be written in the form

$$i = \frac{E}{R}(1 - e^{-t/\tau}) \tag{10.9}$$

In order to determine the value of the current τ seconds after the switch is closed, let $t = \tau$ in equation 10.9

$$(i)_{t=\tau} = \frac{E}{R}(1 - e^{-1}) = \frac{E}{R}(1 - 0.368) = 0.632E/R$$

that is, the current reaches 63.2 per cent of its final value in this period of time. Also, in the same length of time, v_R has risen to 63.2 per cent of its final value, and v_L has fallen by 63.2 per cent of its initial value to $0.368E$.

Decay of current in an inductive circuit

Suppose that switch S in figure 10.3a is closed when $t = 0$, when the inductor is carrying current I. At this instant of time the e.m.f. applied to the LR circuit is reduced to zero, and $v_R + v_L = 0$, or

$$Ri + L\frac{di}{dt} = 0 \tag{10.10}$$

hence

$$Ri = -L\frac{di}{dt} \qquad \text{or} \qquad \frac{R}{L}dt = -\frac{di}{i}$$

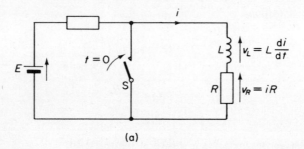

(a)

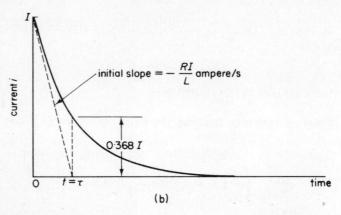

(b)

Figure 10.3   Decay of current in an inductive circuit

whence

$$\int \frac{R}{L}\, dt = -\int \frac{di}{i}$$

therefore

$$\frac{Rt}{L} = -\ln i + \ln A \qquad (10.11)$$

where A is the constant of integration. At the instant that the switch is closed $(t = 0)$, the circuit current is I. Substituting the initial conditions into equation 10.11 gives

$$0 = -\ln I + \ln A$$

or

$$A = I$$

therefore

$$\frac{Rt}{L} = -\ln i + \ln I = \ln (I/i)$$

hence

$$e^{R t/L} = I/i$$

therefore

$$i = Ie^{-Rt/L} \qquad (10.12)$$

$$= Ie^{-t/\tau} \qquad (10.13)$$

The general shape of this curve is shown in figure 10.3b.

Final value of current   The final value of current is attained after an infinite length of time, when

$$(i)_{t=\infty} = Ie^{-\infty} = 0$$

That is, the steady value of circuit current is zero.

Initial rate of fall of current   Inserting the initial value of $i\,(= I)$ into equation 10.10 yields

$$\left(\frac{di}{dt} \right)_{t=0} = -\frac{R}{L}I = -\frac{I}{\tau} \quad \text{A/s} \qquad (10.14)$$

If this rate of fall of current were to continue, the current would reach zero value in τ seconds.

Circuit current when $t = \tau$ Substituting $t = \tau$ into equation 10.13 yields

$$(i)_{t\,=\,\tau} = Ie^{-1} = 0.368\,I$$

that is, the current falls to 36.8 per cent of its initial value in τ seconds.

Example 10.1

A voltage pulse of amplitude 5 V and duration 5 ms is applied to a relay coil of inductance 0.1 H and resistance 100 Ω. The relay contacts close when the coil current is 40 mA and open when it is 15 mA. Determine (a) the time delay before the contacts close and (b) the length of time the contacts remain closed. Assume that the current in the coil is initially zero and that the source resistance is zero.

Solution

The steady-state current in the coil after the switch-on transient has decayed is

$$I = E/R = 5/100 = 0.05 \text{ A} = 50 \text{ mA}$$

The waveform of the current in the coil is shown in figure 10.4 in which

t_1 = time delay before the contacts close

t_2 = time delay for the contacts to open after the applied
         voltage has been reduced to zero

t_3 = length of time the contacts remain closed = 5 ms $- t_1 + t_2$

The circuit time constant is

$$\tau = L/R = 0.1/100 = 0.001 \text{ s}$$

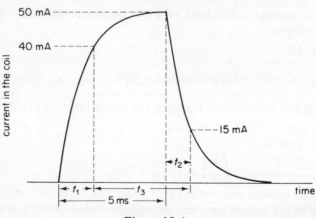

Figure 10.4

(a) *Determination of* t_1

The relay current after t_1 is 40 mA hence, from equation 10.9

$$i = \frac{E}{R}(1 - e^{t_1/\tau})$$

or

$$40 \times 10^{-3} = 50 \times 10^{-3}(1 - e^{-t_1/0.001})$$

Solving for t_1 yields

$$t_1 = \tau \ln 5 = 0.001 \times 1.6094 = 1.6094 \times 10^{-3} \text{ s} = 1.6094 \text{ ms}$$

(b) *Determination of* t_2

Equation 10.13 describes the fall of the current in the circuit, in which $I = 50$ mA, $i = 15$ mA and $t = t_2$, hence

$$15 \times 10^{-3} = 50 \times 10^{-3} \times e^{-t_2/\tau}$$

or

$$t_2 = \tau \ln(50/15) = 10^{-3} \times 1.2039 \times 10^{-3} \text{ s}$$

$$= 1.2039 \text{ ms}$$

(c) *Determination of* t_3

From the expression given above

$$t_3 = 5 - t_1 + t_2 = 5 - 1.6094 + 1.2039 \approx 4.6 \text{ ms}$$

10.2 Sketching Exponential Curves

A method of sketching the general shape of the curves described by equations 10.4 and 10.13 is by evaluating the co-ordinates of a number of points on the curves. The results in table 10.1 are calculated for the equation $i = (E/R)(1 - e^{-t/\tau})$.

Table 10.1

Value of i	Time taken to reach i	Point on figure 10.5a
0.5 E/R	0.7 τ	A
0.75 E/R	1.4 τ	B
0.875 E/R	2.1 τ	C
0.9375 E/R	2.8 τ	D

The resulting curve is sketched in figure 10.5a, and it can be seen that in the first time interval of 0.7 τ the current rises by a value which is one-half of the difference between the initial value at the commencement of the time interval and the final

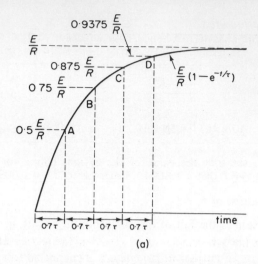

(a)

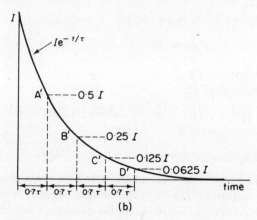

(b)

Figure 10.5   Sketching exponential curves

value (that is, E/R). In the second $0.7\,\tau$ time interval the current rises by one-half of the difference between the value of the current at the beginning of the interval ($0.5\,E/R$) and the final value, that is, by one-half of $(1 - 0.5)E/R = 0.25\,E/R$ (point B on the curve). During the third $0.7\,\tau$ time interval the current again rises by one-half of the difference between the value of the current at the commencement of the period ($0.75\,E/R$) and the final value, that is, by one-half of $(1 - 0.75)\,E/R = 0.125E/R$, to point C on the curve. This relationship continues during the remainder of the curve.

The curve for the decay of current, $i = Ie^{-t/\tau}$, can be plotted from the figures in table 10.2.

Table 10.2

Value of i	Time taken to reach i	Point on figure 10.5b
$0.5I$	$0.7\,\tau$	A'
$0.25I$	$1.4\,\tau$	B'
$0.125I$	$2.1\,\tau$	C'
$0.0625I$	$2.8\,\tau$	D'

From the table, we see that the value of the current halves for each $0.7\,\tau$ time interval.

Example 10.2

An inductive circuit has a resistance of $10\,\Omega$ and an inductance of 1 H. Sketch the curve showing the rise of current in the coil if it is connected to a 20 V d.c. supply of zero source resistance. After steady-state operating conditions the supply is disconnected and, simultaneously, the coil is short-circuited. Sketch the curve showing the decay of current in the circuit.

Solution

The steady-state circuit current is

$$I = E/R = 20/10 = 2 \text{ A}$$

and the circuit time constant is

$$\tau = L/R = 1/10 = 0.1 \text{ s} = 100 \text{ ms}$$

From the preceding work, the current will have risen from zero to 1 A in 70 ms, to 1.5 A in 140 ms, and to 1.75 A in 210 ms, and so on. The curve showing the rise of current is shown in figure 10.6. Also, after the coil has been disconnected from the

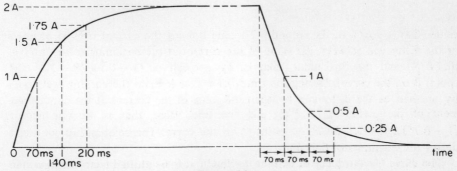

Figure 10.6

supply and a short-circuit applied to it, the current falls to 1 A in 70 ms, to 0.5 A after another 70 ms, etc.

10.3 Rise-time, Fall-time and Settling-time of Exponential Curves

The *rise-time*, t_r, of an exponential curve is the time taken for the waveform to rise from 10 per cent to 90 per cent of the final value, and is illustrated in figure 10.7. Hence

$$t_r = t_2 - t_1$$

At time t_1 the circuit current is $0.1I$, hence from equation 10.9

$$0.1I = I(1 - e^{-t_1/\tau})$$

or

$$t_1 = \tau \ln(1/0.9) \approx 0.1\tau$$

At time t_2 the circuit current is $0.9I$, hence

$$0.9I = I(1 - e^{-t_2/\tau})$$

or

$$t_2 = \tau \ln 10 \approx 2.3\tau$$

Hence

$$\text{rise-time} = t_r = t_2 - t_1 = 2.2\tau \tag{10.15}$$

In example 10.2 the value of τ was 0.1 s, and in that case $t_r = 0.22$ s.

The *settling-time*, t_s, is the time taken for the transient current to die away, and is assumed to have occurred when the value of the current in waveform in figure

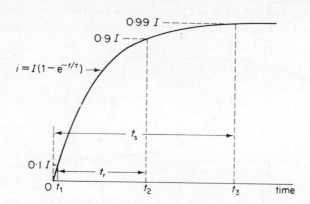

Figure 10.7   Determination of the rise-time and the settling-time of an exponential curve

10.7 has reached 99 per cent of its final value. Its value is equal to t_3 in the figure, and is calculated as follows

$$0.99\,I = I\,(1 - e^{-t_3/\tau})$$

or

$$t_s = t_3 = \tau \ln 100 \approx 4.6\tau \qquad (10.16)$$

In the case of a circuit with a time constant of 0.1 s, the settling-time is 0.46 s.

The *fall-time*, t_f, of an exponential curve is the time taken for it to fall in value from 90 per cent to 10 per cent of its initial value. This is shown as the time interval t_f in figure 10.8, for the decay of current in an inductive circuit. The values of t_4 and t_5 on figure 10.8 are calculated as follows.

When $t = t_4$, $i = 0.9I$, hence from equation 10.13

$$0.9I = Ie^{-t_4/\tau}$$

or

$$t_4 = \tau \ln (1/0.9) \approx 0.1\,\tau$$

and at t_5, $i = 0.1I$, when

$$0.1I = Ie^{-t_5/\tau}$$

or

$$t_5 = \tau \ln 10 \approx 2.3\,\tau$$

hence

$$\text{fall-time} = t_f = t_5 - t_4 = 2.2\tau \qquad (10.17)$$

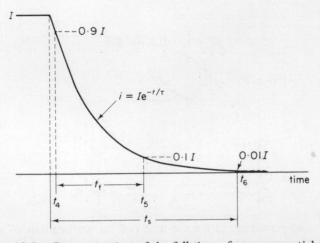

Figure 10.8 Determination of the fall-time of an exponential curve

The *settling-time, t_s,* of figure 10.8 is the time taken for the current to fall to one per cent of its initial value, that is, at time t_6 when

$$0.01I = Ie^{-t_6/\tau}$$

or

$$t_f = t_6 = \tau \ln 100 \approx 4.6\tau \qquad (10.18)$$

10.4 Transients in *RC* Series Circuits

Charging current

If the capacitor in figure 10.9 is initially uncharged, then, when the switch is closed at $t = 0$, the initial rush of charging current is restricted only by the value of resistor R. As a result, the value of the initial charging current is E/R amperes. As the charge stored by the capacitor builds up, so the voltage v_C between its terminals increases and the charging current decays in value. When the capacitor is fully charged the voltage between its terminals is equal to E and the magnitude of the current in the circuit is zero. The equation for the current in the transient period is derived below.

From the work in chapter 4, the capacitor current is given by

$$i = C\frac{dv_C}{dt}$$

Applying Kirchhoff's first law to figure 10.9 gives

$$E = iR + v_C = RC\frac{dv_C}{dt} + v_C$$

or

$$(E - v_C)\,dt = RC\,dv_C$$

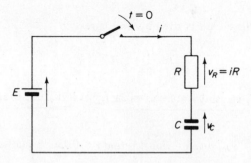

Figure 10.9   *RC* series circuit

hence

$$\frac{dt}{RC} = \frac{dv_C}{(E - v_C)} \qquad (10.19)$$

Integrating both sides of equation 10.19 gives

$$\frac{t}{RC} = -\ln (E - v_C) + \ln A \qquad (10.20)$$

where A is the constant of integration, whose value is determined by inserting the initial conditions in the circuit into the equation. These conditions are that the capacitor is initially uncharged, that is $v_C = 0$ when $t = 0$. Substituting the initial conditions into equation 10.20 gives

$$0 = -\ln E + \ln A$$

hence

$$\frac{t}{RC} = -\ln (E - v_C) + \ln E = \ln\left(\frac{E}{E - v_C}\right)$$

or

$$\frac{E}{E - v_C} = e^{t/RC}$$

whence

$$v_C = E(1 - e^{-t/RC}) \qquad (10.21)$$

$$= E(1 - e^{-t/\tau}) \qquad (10.22)$$

where $\tau = RC$, and is the time constant of the circuit.

The voltage across the resistance is

$$v_R = E - v_C = Ee^{-t/\tau} \qquad (10.23)$$

and the current in the circuit is given by the expression

$$i = \frac{v_R}{R} = \frac{E}{R}e^{-t/\tau} \qquad (10.24)$$

The curves for v_C, v_R and i are shown in figures 10.10a, b and c, respectively.

Initial values of v_C, v_R, and i   Substituting $t = 0$ into equation 10.21 yields the initial value of v_C.

$$(v_C)_{t=0} = E(1 - e^{-0}) = 0$$

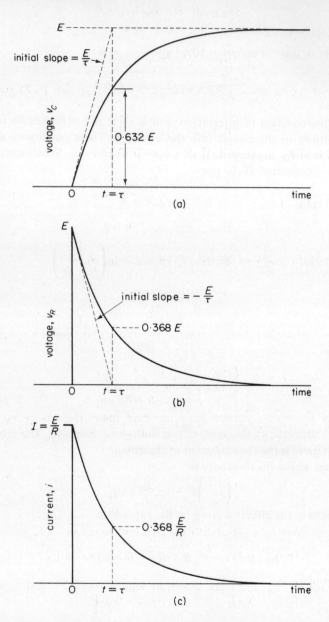

Figure 10.10   Transients during the charging period of an *RC* circuit

and the initial value of v_R is

$$(v_R)_{t=0} = E - (v_C)_{t=0} = E$$

and the initial current is

$$(i)_{t=0} = (v_R)_{t=0}/R = E/R$$

Final values of v_C, v_R and i   Substituting $t = \infty$ into equation 10.21 gives the final value of v_C as

$$(v_C)_{t=\infty} = E(1 - e^{-\infty}) = E$$

The final value of v_R is, therefore

$$(v_R)_{t=\infty} = E - (v_C)_{t=\infty} = 0$$

and the final value of i is

$$(i)_{t=\infty} = (v_R)_{t=\infty}/R = 0$$

Initial slopes of the transient curves   From equation 10.19

$$\frac{dv_C}{dt} = \frac{E - v_C}{RC}$$

Inserting the initial conditions, that is, $v_C = 0$ at $t = 0$ into the above equation gives

$$\left(\frac{dv_C}{dt}\right)_{t=0} = \frac{E}{RC} = \frac{E}{\tau} \quad \text{V/s} \tag{10.25}$$

Clearly, since the supply voltage has a constant value, the voltage v_R across the resistor must decrease at the same rate as voltage v_C increases. The initial rate of change of v_R is, therefore, given by the expression

$$\left(\frac{dv_R}{dt}\right)_{t=0} = -\frac{E}{\tau} \quad \text{V/s}$$

Circuit voltages when $t = \tau$ Substituting $t = \tau$ into equation 10.22 yields

$$(v_C)_{t=\tau} = E(1 - e^{-1}) = E(1 - 0.368) = 0.632\,E$$

and the value of v_R is

$$(v_R)_{t=\tau} = E - (v_C)_{t=\tau} = 0.368\,E$$

Rise-time of v_C   Since v_C follows an exponential curve of the general type in figure 10.7, the rise-time of the curve is

$$\text{rise-time} = t_r = 2.2\tau = 2.2RC \quad \text{seconds} \tag{10.26}$$

Fall-time of v_R The graph of v_R follows the general pattern of the curve in figure 10.8, hence the fall-time is

$$\text{fall-time} = t_f = 2.2\tau = 2.2RC \quad \text{seconds} \tag{10.27}$$

Settling-time of v_C *and* v_R Following the work in section 10.3, the settling-time for the curves in figure 10.10 is

$$\text{settling-time} = t_s = 4.6\tau = 4.6RC \quad \text{seconds} \tag{10.28}$$

Discharge current

Let us assume that a capacitor has been charged to voltage E in the manner described above, and it is re-connected into the circuit in figure 10.11. When the switch is closed, the capacitor discharges its energy into the circuit and current flows *out* of the upper plate and *into* the lower plate.

The convention adopted here is to assume that the positive direction of flow of current is that which causes the capacitor to be charged. Thus, in figure 10.11a, we show the circuit current flowing into the upper plate. In the analysis which follows it will be seen that the resulting value of i is negative (see equation 10.34 below), clearly indicating that the actual direction of flow of current is opposite to the one we have chosen.

Since the e.m.f. acting around the loop to charge the capacitor is zero, then

$$iR + v_C = 0 \tag{10.29}$$

and, since $i = C \, dv_C/dt$, then

$$RC \, dv_C/dt = -v_C \tag{10.30}$$

or

$$\frac{dt}{RC} = -\frac{dv_C}{v_C}$$

Integrating both sides of the above equation yields

$$\frac{t}{RC} = -\ln v_C + \ln A \tag{10.31}$$

The constant of integration, A, is determined by inserting the initial conditions into equation 10.31, that is, $v_C = E$ when $t = 0$.

$$0 = -\ln E + \ln A$$

or

$$A = E$$

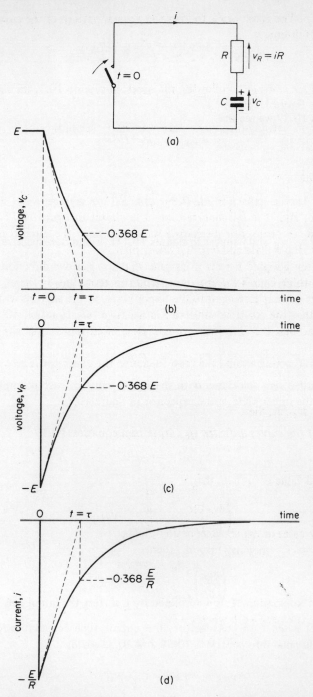

Figure 10.11 Transients in an *RC* circuit during the discharge period

therefore

$$\frac{t}{RC} = -\ln v_C + \ln E = \ln (E/v_C)$$

whence

$$v_C = Ee^{-t/RC} \tag{10.32}$$

From equation 10.29 we see that

$$v_R = -v_C = -Ee^{-t/RC} \tag{10.33}$$

also

$$i = v_R/R = -\frac{E}{R}e^{-t/RC} \tag{10.34}$$

The curves for v_C, v_R and i appear in figures 10.11b, c and d, respectively.

Initial values   Inserting $t = 0$ in equations 10.32 to 10.34 gives

$$(v_C)_{t=0} = E$$

$$(v_R)_{t=0} = -E$$

$$(i)_{t=0} = -E/R$$

Note The negative sign associated with the initial value of current implies that the current *leaves* the upper plate of the capacitor in figure 10.11.

Initial slopes of the curves in figure 10.11   From equation 10.30

$$dv_C/dt = -v_C/RC$$

Since the initial value of v_C is E, then

$$\left(\frac{dv_C}{dt}\right)_{t=0} = -\frac{E}{RC} \quad \text{V/s}$$

Also, since $v_R = -v_C$, then $dv_R/dv_C/dt$, then

$$\left(\frac{dv_R}{dt}\right)_{t=0} = \frac{E}{RC} \quad \text{V/s}$$

Circuit voltages when $t = \tau$ As before, the circuit time constant is RC seconds; substituting this value into equations 10.32 and 10.33 yields

$$(v_C)_{t=\tau} = Ee^{-1} = 0.368 E$$

$$(v_R)_{t=\tau} = -Ee^{-1} = -0.368 E$$

Final values of voltage and current   Putting $t = \infty$ into equations 10.32 to 10.34 gives

$$(v_C)_{t=\infty} = Ee^{-\infty} = 0$$

$$(v_R)_{t=\infty} = -Ee^{-\infty} = 0$$

$$(i)_{t=\infty} = \frac{E}{R}e^{-\infty} = 0$$

Fall-time and settling-time   The fall-times of v_C and of v_R are both given by the expression

$$t_f = 2.2\tau$$

and the settling-time of all the waveforms in figure 10.11 is

$$t_s = 4.6\tau$$

10.5 Sawtooth Waveform Generator or Timebase Generator

The basis of many electronic sawtooth waveform generators is an RC series circuit energised by a unidirectional supply voltage, whose capacitor can be discharged very rapidly at regular intervals of time by means of an electronic switch. One such circuit is shown in figure 10.12a, in which a voltage-sensitive electronic switch, D, is connected across the capacitor. When the voltage across the capacitor reaches voltage V_{BR} at time t_1 in figure 10.12b, the switch closes and short-circuits the capacitor, thereby discharging it. When the capacitor is fully discharged, the switch opens and the capacitor begins to charge once more. This cycle is repeated indefinitely. Electronic devices used as voltage-sensitive switches include diacs

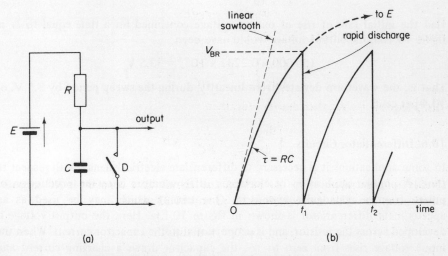

Figure 10.12   A basic form of sawtooth generator

(bi-directional breakdown diodes), unijunction transistors and neon tubes. The circuit is known as a *sawtooth generator* because of the shape of the output voltage waveform. It is also known as a *timebase generator* since it can be used to provide an X-deflection waveform (the timebase waveform) of a cathode-ray oscilloscope.

Ideally, timebase waveforms have a linear slope. The waveform in figure 10.12b follows an exponential curve and is not linear but, provided that V_{BR} is very much less in value than E, then the waveform approximates to a straight line.

If in figure 10.12a, $E = 150$ V, $V_{BR} = 30$ V, $R = 100$ kΩ, and $C = 0.01$ μF then, from the equation for the voltage across the capacitor, equation 10.21

$$v_C = E(1 - e^{-t/\tau})$$

When $t = t_1$ in the above equation, then $v_C = V_{BR}$. Since the circuit time constant is ($100 \times 10^3 \times 0.01 \times 10^{-6} = 10^{-3}$ s), then

$$30 = 150(1 - e^{-t_1/10^{-3}})$$

or

$$e^{-t_1/10^{-3}} = 0.8$$

whence

$$t_1 = 10^{-3}\ln(1/0.8) = 0.2231 \times 10^{-3} \text{ s} = 0.2231 \text{ ms}$$

From the above calculation, the periodic time of the waveform is 0.2231 ms. The deviation of the waveform from linearity during the 'sweep' time can be determined from our knowledge of the initial slope of the waveform which is, from equation 10.25

$$\frac{E}{\tau} = \frac{150}{10^{-3}} = 150\,000 \text{ V/s}$$

Had the initial rate of rise of output voltage continued for a time equal to t_1 in figure 10.12b, the output voltage would have been

$$150\,000 \times 0.2231 \times 10^{-3} = 33.5 \text{ V}$$

That is, the waveform deviates from linearity during the sweep period by 3.5 V, or 10.4 per cent.

10.6 Differentiator Circuits

In some applications it is necessary to differentiate electrical signals with respect to time. A popular application of electronic differentiators is to the production of pulses from a rectangular waveform. One circuit which may be used as an approximate differentiator is shown in figure 10.13a. Here the output voltage is developed across the resistor, and is proportional to the capacitor current. When the input voltage rises from zero to $+E$, the capacitor draws a charging current and causes the upper terminal to become positive for the period of time the capacitor

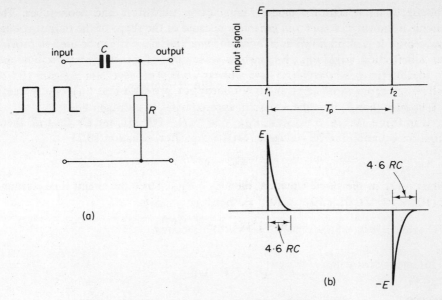

Figure 10.13   Approximate differentiator circuit

charges. When the input voltage falls from $+E$ to zero, it is equivalent to applying a short-circuit to the input terminals of the circuit, in the manner of figure 10.11a, thereby discharging the capacitor.

It was shown earlier that the settling-time for pulses of the type in figure 10.13b is 4.6τ (see equation 10.28). For the output waveform in figure 10.13b to be regarded as the differential of the input waveform, the pulses must be of very short duration. That is

$$4.6\,RC \ll T_\mathrm{p}$$

where T_p is the pulse period of the rectangular wave. If we assume that T_p is about 100 times greater than the product RC, then

$$RC = T_\mathrm{p}/100$$

If T_p is 1 ms, then to differentiate the wave we need an RC circuit with a time constant of 0.01 ms or less; if $C = 0.01\ \mu\mathrm{F}$, the value of R should be 1 kΩ, and if $C = 0.1\ \mu\mathrm{F}$ then $R = 0.1$ kΩ

Summary of essential formulae

RL series circuit: time constant $= \tau = L/R$ seconds

Rise of current: $i = \dfrac{E}{R}(1 - \mathrm{e}^{-t/\tau})$

$$v_R = E(1 - \mathrm{e}^{-t/\tau})$$

$$v_L = Ee^{-t/\tau}$$

final value of current $= E/R$ ampere

initial rate of rise of current $= E/L$ A/s

current when $t = \tau$: $i = 0.632E/R$ ampere

Fall of current: $i = Ie^{-t/\tau}$

$$v_R = IRe^{-t/\tau}$$

final value of current $= 0$

initial rate of fall of current $= -I/\tau$ A/s

current when $t = \tau$; $i = 0.368I$ ampere

RC series circuit: time constant $= \tau = RC$ seconds

Charging current: $i = \dfrac{E}{R}e^{-t/\tau}$

$$v_R = Ee^{-t/\tau}$$
$$v_C = E(1 - e^{-t/\tau})$$

final value of current $= 0$

initial rate of change of $v_C = E/\tau$ V/s

capacitor voltage when $t = \tau$: $v_C = 0.632E$

Discharge current: $i = -\dfrac{E}{R}e^{-t/\tau}$

$$v_R = -Ee^{-t/\tau}$$
$$v_C = Ee^{-t/\tau}$$

final value of current $= 0$

initial rate of change of $v_C = -E/\tau$ V/s

capacitor voltage when $t = \tau$: $v_C = 0.368E$

Exponential curves:

$$\text{rise-time} = t_r = 2.2 \, \tau$$

$$\text{fall-time} = t_f = 2.2 \, \tau$$

$$\text{settling-time} = t_s = 4.6 \, \tau$$

PROBLEMS

10.1 A coil of resistance 100 Ω and inductance 1.0 H is connected to a 100-V d.c. supply. Determine (a) the final value of current in the coil, (b) the initial rate of change of current in the coil, (c) the time constant of the circuit, (d) the value of the current after a time equal to two time constants, (e) the time taken for the current to reach one-half of its final value and (f) the current in the coil after 0.03 s.
[(a) 1 A; (b) 100 A/s; (c) 0.01 s; (d) 0.865 A; (e) 0.0069 s; (f) 0.95 A]

10.2 A coil of inductance 1.5 H and resistance 100 Ω is energised by a d.c. supply. When the supply is disconnected it is arranged so that the current decays through a circuit of resistance R Ω. If the current falls to 60 per cent of its initial value in 2 ms, determine the value of R.
[1370 Ω]

10.3 For figure 10.14, determine (a) the value of i_1 and i_2 1.4 s after closing switch S and (b) the final value of i_2.

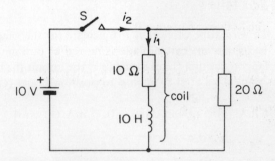

Figure 10.14

Switch S is opened after the initial transient has decayed; determine at the instant the switch contacts are opened (c) the current in the 20-Ω resistor and (d) the p.d. across the switch contacts.
[(a) i_1 = 0.753 A, i_2 = 1.253 A; (b) 1.5 A; (c) 1 A; (d) 30 V]

10.4 Determine the time constant of an *RC* series circuit containing a 0.01-μF capacitor and a 1-MΩ resistor.
The above circuit is connected to a 100-V d.c. supply. Determine (a) the initial value of the charging current, (b) the initial rate of rise of voltage across the capacitor, (c) the voltage across the capacitor 0.03 s after the supply is connected and (d) the charge ultimately stored by the capacitor.
[(a) 100 μA; (b) 10 kV/s; (c) 95.02 V; (d) 50 μC]

10.5   A capacitor of capacitance 10 μF is connected in series with a 250-V volt-meter of resistance 50 kΩ, the circuit being connected to a 200-V supply. Determine after a time interval of 1.0 s (a) the voltage indicated by the voltmeter, (b) the voltage across the capacitor and (c) the current in the circuit. Calculate also the charging current when the voltmeter indicates 100 V.
[(a) 27.1 V; (b) 172.9 V; (c) 0.541 mA; 2 mA]

10.6   The blade of the switch in figure 10.15 is switched from position A to position B at $t = 0$, the capacitor being initially discharged. After a time interval of 0.3 s the switch blade is changed to position C.

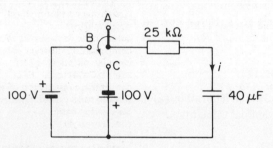

Figure 10.15

Draw to scale the waveshape of current i. Calculate (a) the value of the current in the circuit at the instant the switch blade is moved to position B, (b) its value just prior to switching to position C, (c) its value at the instant the blade changes to position C, (d) its value 0.3 s after switching to position C and (e) the final value of i.
[(a) 4 mA; (b) 2.96 mA; (c) − 1.04 mA; (d) − 0.77 mA; (e) zero]

Index